When Things Start to Think

When Things Start to Think

Neil Gershenfeld

◆ ◆ ◆ ◆ ◆

AN OWL BOOK

HENRY HOLT AND COMPANY ◆ NEW YORK

Henry Holt and Company, LLC
Publishers since 1866
115 West 18th Street
New York, New York 10011

Henry Holt® is a registered trademark
of Henry Holt and Company, LLC.

Published in Canada by Fitzhenry & Whiteside Ltd.,
195 Allstate Parkway, Markham, Ontario L3R 4T8.

Library of Congress Cataloging-in-Publication Data
Gershenfeld, Neil A.
When things start to think / Neil Gershenfeld.
p.   cm.
Includes index.
ISBN 0-8050-5880-X
1. Computers and civilization.   2. Digital electronics.
3. Artificial intelligence.   I. Title.
QA76.9.C66G465   1999   98-30310
303.48′34—dc21   CIP

Henry Holt books are available for special promotions and
premiums. For details contact: Director, Special Markets.

First published in hardcover in 1999 by Henry Holt and Company

First Owl Books Edition 2000

Designed by Victoria Hartman

Printed in the United States of America

1 3 5 7 9 10 8 6 4 2

For Laura
who isn't impressed by gadgets

✦

and for Grace and Eli
who are

Contents

HOW

Preface

I have a vested interest in the future, because I plan on living there. I want to help create one in which machines can meet the needs of people, rather than the other way around. As more and more gadgets demand our attention, the promises of the Digital Revolution start to sound more like those of a disinformation campaign. A counterrevolution seeks to protect our freedom to not be wired. But rather than ask whether reading is better done with a book or a computer, I want to show how to merge the best features of each.

This modest aim has radical implications, and regrettably little precedent. Putting ink on paper, and data on a CD-ROM, are separated by a divide between the old analog world of atoms and the new digital world of bits. The existing organization of industry and academia, of scientific research and product development, serves to enforce rather than dismantle that distinction. I've spent my life being gently and not-so-gently steered away from the equally great problems and opportunities that lie neglected at the boundary between the content of information and its physical representation.

In high school I fell in love . . . with the machine shop. This was the one class in which I showed some promise. Then the curriculum split, with the college-bound students sitting in classrooms, and the vocational students going off to a trade school where they could use an even better machine shop, and make circuits, and assemble engines, and build houses. There was no competition—*that's* what I wanted to do. With some effort I was persuaded to stick with the college track, but I couldn't understand why making things had to be separated from learning about things. I still don't. At Bell Labs I was threatened with a grievance for daring to leave my lab and approach the machine tools in the unionized shop, which was intended for the trade school graduates rather than the college graduates.

As an undergraduate I migrated from studying philosophy in order to ask deep questions about the universe, to studying physics in order to answer deep questions about the universe, and eventually ended up back in the machine shop in the basement of the engineering building rediscovering the joys of manipulating things rather than ideas. In physics grad school I figured out that theorists weren't allowed in the machine shop, but experimentalists were allowed to do theory, so the only way to do both was to claim to be an experimentalist.

I was officially a physicist by the time I visited the Media Lab to develop sensors to interface Yo-Yo Ma's cello to a computer. This prompted some people to ask me a question I never understood: "That's fine, but is it physics?" I could answer it, explaining where the traditional physics occurred. But I didn't want to, much preferring to discuss the remarkable consequences of connecting Yo-Yo to a computer.

When I arrived at the Media Lab I didn't see it as a place to do "real" science, but it was only there that I was able to bring together these disconnected pieces. This was finally a place where I could set up a lab that put machine tools, theoretical physicists, and musicians in the same room at the same time. As it began filling with students and sponsors, I was delighted to find that I was not alone in struggling to bridge rather than bound the digital and physical worlds.

Emerging from this effort is a vision of a future that is much more accessible, connected, expressive, and responsive, as well as many of the ingredients needed to realize the vision. This book tells that story. I feel safe in making these predictions about the future because they're really observations about the present. In the laboratory and in product development pipelines, information is moving out of traditional computers and into the world around us, a change that is much more significant than the arrival of multimedia or the Internet because it touches on so much more of human experience.

Much of the book revolves around the Media Lab, because much of my life does. Although this book is a personal statement, it owes a deep debt to the dear community of colleagues, students, and sponsors there who have shaped these ideas. I describe global changes that still have too few local environments that are as supportive of bringing bits and atoms together. My hope is that the book's focus will come to appear to be parochial, as the Media Lab's role becomes less unique.

This book reflects recurring conversations that I've had with ordinary people about how the world around them does and will work, with industrial managers trying to navigate through coming business threats to reach these new opportunities, and with researchers struggling with emerging questions in new areas along with declining interest and support in old ones. I've written for all three because I've never been able to separate these parts of myself.

I will have succeeded if a shoe computer comes to be seen as a great idea and not just a joke, if it becomes natural to recognize that people and things have relative rights that are now routinely infringed, if computers disappear and the world becomes our interface.

When Things Start to Think

The Digital Evolution

To a species that seeks to communicate, offering instantaneous global connectivity is like wiring the pleasure center of a rat's brain to a bar that the rat then presses over and over until it drops from exhaustion. I hit bottom when I found myself in Corsica, at the beach, in a pay phone booth, connecting my laptop with an acoustic modem coupler to take care of some perfectly mundane e-mail. That's when I threw away my pager that delivered e-mail to me as I traveled, disconnected my cell phone that let my laptop download e-mail anywhere, and began answering e-mail just once a day. These technologies were like the rat's pleasure bar, just capable enough to provide instant communication gratification, but not intelligent enough to help me manage that communication.

The Digital Revolution has promised a future of boundless opportunity accessed at the speed of light, of a globally wired world, of unlimited entertainment and education within everyone's reach. The digital reality is something less than that.

The World Wide Web touches the rather limited subset of human experience spent sitting alone staring at a screen. The way we browse the Web, clicking with a mouse, is like what a child does sitting in a shopping cart at a supermarket, pointing at things of interest, perpetually straining to reach treats that are just out of reach. Children at play present a much more compelling metaphor for interacting with information, using all of their senses to explore and create, in groups as well as alone. Anyone who was ever a child understands these skills, but computers don't.

The march of technical progress threatens to turn the simple pleasure of reading a nicely bound book, or writing with a fountain pen, into an antitechnical act of defiance. To keep up, your book should be on a CD-ROM, and your writing should be done with a stylus on a computer's input tablet. But the reality is that books or fountain pens really do perform their intended jobs better than their digital descendants.

This doesn't mean we should throw out our computers; it means we should expect much more from them. As radical as it may sound, it actually is increasingly possible to ask that new technology work as well as what it presumes to replace. We've only recently been able to explain why the printing on a sheet of paper looks so much better than the same text on a computer screen, and are just beginning to glimpse how we can use electronic inks to turn paper itself into a display so that the contents of a book can change.

Unless this challenge is taken seriously, the connected digital world will remain full of barriers. A computer with a keyboard and mouse can be used by only one person at a time, helping you communicate with someone on the other side of the world but not with someone in the same room. Inexpensive computers still cost as much as a used car, and are much more difficult to understand how to operate, dividing society into an information-rich upper class and an information-poor underclass. A compact disc player faithfully reproduces the artistic creations of a very small group of people and turns everyone else into passive consumers. We live in a three-dimensional world, but displays and printers restrict information to two-dimensional surfaces. A desktop computer requires a desk, and a laptop computer requires a lap, forcing you to sit still. Either you can take a walk, or you can use a computer.

These problems can all be fixed by dismantling the real barrier, the one between digital information and our physical world. We're made out of atoms, and will continue to be for the foreseeable future. All of the bits in the world are of no use unless they can meet us out here on our terms. The very notion that computers can create a virtual reality requires an awkward linguistic construction to refer to "real" reality, what's left outside the computer. That's backward. Rather than replace our world, we should first look to machines to enhance it.

The Digital Revolution is an incomplete story. There is a disconnect between the breathless pronouncements of cyber gurus and the experience of ordinary people left perpetually upgrading hardware to meet the demands of new software, or wondering where their files have gone, or trying to understand why they can't connect to the network. The revolution so far has been for the computers, not the people.

Digital data of all kinds, whether an e-mail message or a movie, is encoded as a string of 0's and 1's because of a remarkable discovery by Claude Shannon and John von Neumann in the 1940s. Prior to their work, it was obvious that engineered systems degraded with time and use. A tape recording sounds worse after it is duplicated, a photocopy is less satisfactory than an original, a telephone call becomes more garbled the farther it has to travel. They showed that this is not so for a digital representation. Errors still occur in digital systems, but instead of continuously degrading the performance there is a threshold below which errors can be corrected with near certainty. This means that you can send a message to the next room, or to the next planet, and be confident that it will arrive in the form in which you sent it. In fact, our understanding of how to correct errors has improved so quickly over time that it has enabled deep-space probes to continue sending data at the same rate even though their signals have been growing steadily weaker.

The same digital error correction argument applies to manipulating and storing information. A computer can be made out of imperfect components yet faithfully execute a sequence of instructions for as long as is needed. Data can be saved to a medium that is sure to degrade, but be kept in perpetuity as long as it is periodically copied with error correction. Although no one knows how long a single CD will last, our society can leave a permanent legacy that will outlast any stone tablet as long as someone or something is around to do a bit of regular housekeeping.

In the 1980s at centers such as MIT's Media Lab, people realized that the implications of a digital representation go far beyond reliability: content can transcend its physical representation. At the time ruinous battles were being fought over formats for electronic media such as videotapes (Sony's Betamax vs. Matsushita's VHS) and high-definition television (the United States vs. Europe vs. Japan). These were analog formats, embodied in the custom circuitry needed to decode them. Because hardware sales were tied to the format, control over the standards was seen as the path to economic salvation.

These heated arguments missed the simple observation that a stream of digital data could equally well represent a video, or some text, or a computer program. Rather than decide in advance which format to use, data can be sent with the instructions for a computer to decode them so that anyone is free to define a new standard without needing new hardware. The consumers who were supposed to be buying high-definition television sets from the national champions are instead buying PCs to access the Web.

The protocols of the Internet now serve to eliminate rather than enforce divisions among types of media. Computers can be connected without regard to who made them or where they are, and information can be connected without needing to artificially separate sights and sounds, content and context. The one minor remaining incompatibility is with people.

Billions of dollars are spent developing powerful processors that are put into dumb boxes that have changed little from the earliest days of computing. Yet for so many people and so many applications, the limitation is the box, not the processor. Most computers are nearly blind, deaf, and dumb. These inert machines channel the richness of human communication through a keyboard and a mouse. The speed of the computer is increasingly much less of a concern than the difficulty in telling it what you want it to do, or in understanding what it has done, or in using it where you want to go rather than where it can go.

More transistors are being put on a chip, more lines of code are being added to programs, more computers are appearing on the desks in an office, but computers are not getting easier to use and workers are not becoming more productive. An army of engineers is continuing to solve these legacy problems from the early days of computing, when what's needed is better integration between computers and the rest of the world.

The essential division in the industry between hardware and software represents the organization of computing from the system designer's point of view, not the user's. In successful mature technologies it's not possible to isolate the form and the function. The logical design and the mechanical design of a pen or a piano bind their mechanism with their user interface so closely that it's possible to use them without thinking of them as technology, or even thinking of them at all.

Invisibility is the missing goal in computing. Information technology is at an awkward developmental stage where it is adept at communicating its needs and those of other people, but not yet able to anticipate yours. From here we can either unplug everything and go back to an agrarian society—an intriguing but unrealistic option—or we can bring so much technology so close to people that it can finally disappear. Beyond seeking to make computers ubiquitous, we should try to make them unobtrusive.

A VCR insistently flashing 12:00 is annoying, but notice that it doesn't know that it is a VCR, that the job of a VCR is to tell the time rather than ask it, that there are atomic clocks available on the Internet that can give it the exact time, or even that you might have left the room and have no interest in what it thinks the time is. As we increasingly cohabit with machines, we are doomed to be frustrated by our creations if they lack the rudimentary abilities we take for granted—having an identity, knowing something about our environment, and being able to communicate. In return, these machines need to be designed with the presumption that it is their job to do what we want, not the converse. Like my computer, my children started life by insisting that I provide their inputs and collect their outputs, but unlike my computer they are now learning how to do these things for themselves.

Fixing the division of labor between people and machines depends on understanding what each does best. One day I came into my lab and found all my students clustered around a PC. From the looks on their faces I could tell that this was one of those times when the world had changed. This was when it had just become possible to add a camera and microphone to a computer and see similarly equipped people elsewhere. There was a reverential awe as live faces would pop up on the screen. An early Internet cartoon showed two dogs typing at a computer, one commenting to the other that the great thing about the Internet was that no one could tell that they were dogs. Now we could.

And the dogs were right. The next day the system was turned off, and it hasn't been used since. The ghostly faces on the screen couldn't compete with the resolution, refresh rate, three-dimensionality, color fidelity, and relevance of the people outside of the screen. My students found that the people around them were generally much more interesting than the ones on the screen.

It's no accident that one of the first Web celebrities was a thing, not a person. A camera trained on the coffeepot in the Computer Science department at Cambridge University was connected to a Web server that could show you the state of the coffeepot, timely information of great value (if you happen to be in the department and want a cup of coffee).

Even better would be for the coffee machine to check when you want coffee, rather than the other way around. If your coffee cup could measure the temperature and volume of your coffee, and relay that information to a computer that kept track of how much coffee you drank at what times, it could do a pretty good job of guessing when you would be coming by for a refill. This does not require sticking a PC into the cup; it is possible to embed simple materials in it that can be sensed from a distance. The data handling is also simple; a lifetime record of coffee drinking is dwarfed by the amount of data in a few moments of video. And it does not require a breakthrough in artificial intelligence to predict coffee consumption.

A grad student in the Media Lab, Joseph Kaye, instrumented our coffee machine and found the unsurprising result that there's a big peak in consumption in the morning after people get in, and another one in the afternoon after lunch. He went further to add electronic tags so that the coffee machine could recognize individual coffee cups and thereby give you the kind of coffee you prefer, along with relevant news retrieved from a server while you wait. No one of these steps is revolutionary, but taken together their implications are. You get what you want—a fresh cup of coffee—without having to attend to the details. The machines are communicating with each other so that you don't have to.

For all of the coverage of the growth of the Internet and the

World Wide Web, a far bigger change is coming as the number of things using the Net dwarfs the number of people. The real promise of connecting computers is to free people, by embedding the means to solve problems in the things around us.

There's some precedent for this kind of organization. The recurring lesson we've learned from the study of biology is that hard problems are solved by the interaction of many simple systems. Biology is very much a bottom-up design that gets updated without a lot of central planning. This does not mean that progress is steady. Far from it; just ask a dinosaur (or a mainframe). Here we come to the real problem with the Digital Revolution. Revolutions break things. Sometimes that's a necessary thing to do, but continuously discarding the old and heralding the new is not a sustainable way to live.

Very few mature technologies become obsolete. Contrary to predictions, radio did not kill off newspapers, because it's not possible to skim through a radio program and flip back and forth to parts of interest. Television similarly has not eliminated radio, because there are plenty of times when our ears are more available than our eyes. CD-ROMs have made little dent in book publishing because of the many virtues of a printed book. Far more interesting than declaring a revolution is to ask how to capture the essence of what works well in the present in order to improve the future.

The real challenge is to figure out how to create systems with many components that can work together and change. We've had a digital revolution; we now need digital evolution. That's what this book is about: what happens when the digital world merges with the physical world, why we've come to the current state-of-the-art (and state-of-the-artless), and how to free bits from the confines of computers. The machines have had their turn; now it's ours.

WHAT

. . . are things that think?

books that can change into other books

✦

musical instruments that help beginners
engage and virtuosi do more

✦

shoes that communicate through
body networks

✦

printers that output working things
instead of static objects

✦

money that contains behavior
as well as value

Bits and Books

The demise of the book has been planned for centuries. This came first by fiat, with bannings and burnings, and more recently by design, with new media promising to make old books obsolete. Today's book-of-the-future is the CD-ROM, offering video and sounds and cross-references to enhance ordinary text. Who would ever want to go back to reading a book that just has words?

Just about everyone. The state of the book business can be seen at the Frankfurt book fair, the annual event where the publishing world gathers to buy and sell books. It dates all the way back to the fifteenth century, shortly after the development of movable metal type created the need for a new kind of information marketplace. Approaching the end of the millennium the fair is flourishing, with hundreds of thousands of books and people on display. The only exception is the short lines and long faces in the hall displaying books on disk. People are voting with their fingers.

You don't need to look beyond the book you're holding to understand why. Think about its specifications. A book:

- boots instantly
- has a high-contrast, high-resolution display

- is viewable from any angle, in bright or dim light
- permits fast random access to any page
- provides instant visual and tactile feedback on the location
- can be easily annotated
- requires no batteries or maintenance
- is robustly packaged

A laptop meets exactly none of those specifications. If the book had been invented after the laptop it would be hailed as a great breakthrough. It's not technophobic to prefer to read a book; it's entirely sensible. The future of computing lies back in a book.

Isn't it curious that a book doesn't need a backlight? A laptop screen requires a power-hungry lamp, and even so isn't legible in bright light or from an angle. Under those same circumstances a book continues to look great, and the reason why is surprisingly interesting.

The light in a laptop starts its journey in a fluorescent tube behind the display. It's guided by a panel that spreads the light across the screen, losing some away from the screen. The light then passes through a polarizing filter that transmits the part of the light wave that is oriented in one direction, while absorbing the rest of it. Next come the control electrodes and then the liquid crystal. This is a fluid that can rotate the orientation of the light wave based on a voltage applied by a transistor, which absorbs still more light. Finally the light passes through color filters and a final polarizer, each taking its cut out. The result is that most of the light is wasted inside the display. The fraction that trickles out must then compete with the ambient light around the display to be seen. And since the light that does emerge must make its way through this obstacle course, it continues on in the same direction straight out, leaving very little to be seen at an angle to the display.

A piece of paper takes a much more sensible approach. The fibers that make up a sheet are actually translucent. Light striking the paper gets bent as it passes through a fiber, and since there are so many fibers this happens many times in many directions. The result of all this scattering is that the light spreads through the paper much like the spread of a drop of ink, eventually leaking back out. And just as the shape of a blob of spreading ink doesn't depend much on the angle at which a pen is held, the light comes out of the paper in all directions, independent of how it arrived at the paper. This phenomenon, called *optical weak localization*, is what makes paper (or a glass of milk) appear to be white. It's a very efficient system for converting light of most any color and orientation into uniform background illumination for text on the page. Crucially, like an aikido master redirecting an incoming attacker, this mechanism takes advantage of the light already in a room rather than trying to overpower it the way a backlight does.

In the Media Lab, as I started spending time with publishing companies as well as computer companies, I was struck by how strange it is to replace paper with displays that are guaranteed to be bulkier, take more power, and look worse. If paper is such a good system, why not continue to use it? The one advantage that a liquid crystal panel has over a sheet of paper is that it can change. Joe Jacobson came to the Media Lab to fix that.

He found that the key to making smart paper is a process called *microencapsulation*. This process grows tiny shells of one material around a core of another one. A familiar success of microencapsulation is carbonless copy paper. It used to be that to make a copy of a receipt when it was written, one had to slide in a sheet of an invariably messy paper that had an ink on it that could be transferred by writing on it, or touching it, or holding it, or rubbing it on your clothes. Now it's possible to get a copy simply by writing on what looks like an ordinary piece of paper. The carbonless copy paper has been coated with small particles containing ink. The act of writing on it applies enough force to break open the shells, releasing the ink onto the sheet below. Simply touching the paper doesn't have enough force to break the shells, which leaves the ink encapsulated where it belongs.

The beauty of microencapsulation is that it's also cheap. There's no need for a tiny assembly line to fill the particles; this is done by a straightforward chemical process. A solution is formed of drops of ink of the appropriate size in another liquid. Then a material is introduced that grows only at the interface between the two liquids. Once it has had time to form a thick enough shell around the droplets, the newly microencapsulated particles can be separated out from the liquid.

The toner used in a printer or copier consists of small particles that get fused to a piece of paper and absorb light. Joe's group has developed a way to make microencapsulated particles about the size of toner (smaller than the thickness of a hair), which contain still smaller particles. The inner particles come in two types, one white and the other black. They also have a different electric charge stored on them. This means that in an electric field all of the white particles will go to one side of the outer shell and the black particles to the other. If the field is reversed, they will change sides. The result looks just like toner, because it essentially is, but it's a kind of toner that can be switched on and off. They call this electronic ink.

The first thing that can be done with e-ink is to cover a sheet with it to make reusable paper. Instead of ending up in a recycling bin, the paper can go back into the printer after it's used. A conventional printer has a complex mechanism to spread toner or ink over a page. Reusable paper starts with the smart toner already on it; the printer just needs a row of electrodes to switch the particles on and off. And since the process is reversible, the paper can go back through the printer to be reprinted over and over again.

Reusable paper is needed because tomorrow's paperless office has turned into today's insatiable consumer of paper, carting in reams of paper to fly through ever-faster printers and just as quickly be thrown out. Reusable paper ends this high-tech deforestation by recycling the paper right in the printer. Instead of filling your trash with newsprint, a newspaper printed on reusable paper can go back into the printer at the end of the day to reemerge printed with the next day's news.

Even better than reusing a sheet of paper is changing it while you watch. There's enough room to drive a microelectronic truck through the thickness of a sheet of paper. This means that the electrodes needed to switch the particles can be moved from the printer to the paper itself. Forming a sandwich of paper, particles, and electrodes creates a new kind of display. But unlike any other display this one works just like printing and therefore can be viewed from most any angle in most any light, and it retains its image when the power is switched off. And even though the original goal of the project was to make displays that would look beautiful but be changed infrequently, the particles can switch quickly enough to approach the rate needed to show the moving images of a video.

Joe's group began developing a printer to put the electrodes onto a sheet of paper to make electronic ink displays and soon realized that they could bring still more capabilities to a piece of paper. Integrated circuits are made by depositing conducting and insulating materials onto a wafer held inside a chamber that has had all the air evacuated so that it doesn't react. The head of his group's remarkable prototype printer accomplishes the same thing under air on a desktop by creating a tiny plasma column, like a miniature lightning bolt. This delivers to the surface the materials needed to print circuit elements such as wires and transistors. They developed the printer to put down display circuits to address individual pixels on the paper, but there's plenty of space compared to a computer chip to add other functions.

The most exciting prospect of all is called radio paper. Since it takes very little energy to switch the electronic ink particles they can be driven by a solar cell, which is essentially a big transistor that can be deposited by the circuit printer. By printing still more transistors, a radio receiver could be integrated with the paper. Now your newspaper doesn't even have to go into the printer at the end of the day. If you leave it out on your coffee table, the light in the room will power its circuits, which will receive a radio signal with the news to update the page. The circuitry can then flip the particles to reprint the page to have the day's news waiting for you whenever you pick up the paper. Bad news for birdcages and yesterday's fish: this would be a newspaper that is never out-of-date. While there's still a lot of work to do to make radio paper a reality, the essential elements have already been shown in the laboratory.

Actively updating a newspaper points to the ultimate application of electronic ink: the universal book. Sheets of paper covered with the microencapsulated particles and electrodes can be assembled into a book, with control electronics embedded in the spine. This would look and feel like any other book, since it's made out of the same ingredients of paper and toner. The only difference is that this book can change. After reading it, you can download new contents through the binding and onto the pages. This can even be done while you read. Pages other than the one you're looking at can be changing, so that all of *War and Peace* could be read in a pamphlet of just a few pages.

The great innovation of Gutenberg and his peers was not the printing press, which was just a converted wine press; it was movable metal type. Before then printing was done from wooden blocks that had to be laboriously engraved. Movable type made it possible for a printer to produce a set of type that could be used to print all books, arranging the letters as needed for each page. Electronic ink takes this idea to its logical conclusion, moving the toner instead of the type. Now a reader can own just one book that can be any book, which can arrange the ink as needed for each page. Instead of going to a library to check out a book, the bits of the book can be downloaded onto the pages of an electronic book.

The electronic book ends the argument over old-fashioned books versus new-fashioned bits by recognizing that both sides have strong technical cases that can be combined. There are deep reasons why the old technologies in a book work so well, and there are new ways to emulate and adapt them. What jumps out the first time that you see a Gutenberg Bible is the glossiness of the ink. It turns out that Gutenberg made his inks by cooking a stew of oil and copper and lead that precipitated out little platelets that act like tiny mirrors, paradoxically reflecting light from a black background. The printing industry is still catching up to the formulation of inks of that sophistication. The same applies to the relatively recent discovery of how optical weak localization helps illuminate a printed page. It's fair to ask that any successor to the book be able to do these things as well.

Drawing on all the resources now at our disposal to catch up to what Gutenberg was doing in the fifteenth century is a worthy challenge. But it's not the real goal; it's an essential warm-up along the path to asking what we can now achieve that Gutenberg could not. Gutenberg did exactly the same thing. He started by printing replicas of illuminated manuscripts; after all, that was where the market was. Then, as printing with movable metal type made it possible to assemble more pages than before, it became necessary to invent page numbering and tables of contents to keep the information accessible. These innovations departed from the past practice of manuscripts that were copied by hand, which were designed to minimize length at all costs. There was an interesting transitional period during which hand-illuminated manuscripts added page numbers to keep up with the fashion, before the bother of copying by hand became too hard to justify and quietly disappeared. Freed from the constraint of duplicating manuscripts, Aldus Manutius in Venice around 1500 then settled on the dimensions of the modern book. It was designed to fit in the saddlebags of traveling scholars, and his Press developed the italics font, to fill pages more efficiently than did fonts that imitated handwriting.

The arrival of electronic books now presents an opportunity to rethink how collections of books are organized into libraries. A library is, of course, much more than a book database. It's a reading temple, a place for serendipitous browsing, a space to be with books, by yourself and with others. My only literary connection with John Updike is that we have the same favorite part of Harvard's great Widener library—a small depression in the stone floor on the way into the stacks. To gain admittance to the stacks you pass through a gauntlet of dour gatekeepers, climb a set of stairs, turn immediately right, and then left. The generations of readers and writers pivoting at this last corner over the years have worn it down, hollowing out a bowl that provides a connection to everyone who has come before. I can't write like Updike, but when it comes to floor abrasion I'm on an equal footing with him.

Once you're past that corner, the bookshelves appear. I have no idea how far they extend, because it's impossible to pass more than a few aisles without stopping to pick up an interesting title, then wondering about a book next to it, then realizing that you simply must spend more time in that section. The same thing happens with the old card catalog. Drawers holding worn cards that appear to date back to the opening of the building, if not the Civil War, speak volumes about what books are popular and what books are not. Flipping through the cards is even more likely than walking down the aisle to turn a simple query into a branching ramble through accumulated wisdom, which lasts until dinner or darkness forces an end.

In the noisy battle between the wired digerati and their analog enemies, there is a common misconception that the pleasure and utility of this kind of browsing must be sacrificed if a computer is used to access information. Too much is made of the inadequacies of transitional technology, while simultaneously too little is expected of it. An on-line card catalog that reduces each entry to a single screenful of flickering green text is incapable of communicating the popularity or context of a book. That's not a failing of electrons or phosphors; it's a poorly designed interface. The wear and tear of a card catalog, or of a library floor, are consequences of engineering decisions about the construction of cards, or floors, or shoes. That those technologies succeed through a combination of intended and unplanned interactions provides valuable guidance for future development, but it doesn't mean that they are the only possible solutions.

There are many things that a card catalog, even Widener's, can't do. Of all major U.S. libraries, Widener is the only one that doesn't shelve its books according to the Library of Congress indexing system. This is because Harvard's system predates that of the Library of Congress; relabeling and reshelving the millions of volumes would be prohibitive. It's much easier to move the bits associated with the books. Once a card catalog becomes electronic, the presentation can be freed from a single ordering defined with great effort by generations of librarians. Books can be shown by date, or location, or author, or discipline. Browsing need not be confined to books that are nearby in any one index; a smarter card catalog can show books that are related in multiple ways. And the card catalog can help provide context for the book by showing other books that have been checked out with it, or perhaps a map of where in the world the book is being read then.

Doing these things requires a visual sophistication that was beyond the means of not only early computers but also their programmers. Letting the programmers alone design the card catalog interface results in a faithful presentation of how the computer represents information internally, rather than how people use it externally. Creating books and libraries requires typographers and architects as well as bookbinders and librarians; the same thing is even more true electronically.

My colleague David Small belongs to a new generation of graphic designers who happen to be fluent programmers. Early fonts were designed to hold ink on the letter forms as well as to help guide the eye; David and his peers are using the interactivity of computers to let letters act and react in ways that are appropriate for the text and the reader. One of my favorite of his design studies presents Shakespeare's works through the medium of a graphical supercomputer connected to a Lego set.

Initially the screen shows a hazy blob that in fact contains all of the words of all of Shakespeare's plays. It's a 3D space that you can fly through, so that you can grab the computer's control knobs and zoom in to find a play, and approach still closer to find the text of a scene beautifully laid out. Rotating the space to peek behind the words reveals supporting annotations. Moving within a play and among plays is done by traveling in this space rather than by pulling books off of a shelf.

Since most of us don't have much experience with flying through words, David adds some familiar guides. In front of the computer is a Lego set, with toy characters drawn from Shakespeare's characters and themes. Selecting one of these physical icons highlights the corresponding parts of Shakespeare's corpus on the screen in the background, showing first the large-scale patterns and then the individual words if you move in for a closer look. Spending time in this environment invites browsing in a way that can't be done if the texts are frozen on pages. A good book uses a linear exposition to convey a nonlinear web of connections between the author and the reader; David's environment comes much closer than a flat book to representing how I think about a Shakespeare play as I read it.

There's still one more benefit of electronic books: access. There's an implicitly elitist undercurrent in defending printed books over electronic books. The collection of the Widener library is a treasure that by design very few people will ever get to see, much less use routinely. Before printing, when books were copied by hand, they were so precious that owning one was a sign of great wealth, usually limited to religious orders rather than individual people. A single book is now so inexpensive that anyone can have one, but private libraries are still owned only by institutions and the wealthy. The budgets of most public libraries restrict their collections to a small subset of the wonders of Widener. Creating an electronic Widener is a heroic task that, even if done imperfectly, will make much more information available to many more people. Not everyone can own David Small's supercomputer, but if the job of a local library becomes one of providing the tools to access information rather than holding the information itself, then it can do more with less.

Electronic books will be able to do everything that printed books can do, but one. They can't replace the primacy of a historical artifact. This is a point that technologists like me frequently miss. Part of the pleasure in seeing a Gutenberg Bible is knowing that Gutenberg held the same object that you are looking at, drawing a connection across the centuries between you and him and everyone else who has come in between. A completely faithful replica that matches the specifications can never convey that.

Some information simply can't be copied digitally. The rarest of the rare books in Harvard's collection is bound in human skin. It is a modern witness to an old practice of people leaving a legacy of a bit of themselves for their successors in the binding of a posthumous book. The unsettling attraction of such a book is that it can never be duplicated.

For people who come to a library to experience the texture and smell and even the pedigree of an old book, there's no point in arguing the merits of electronic ink. On the other hand, there's no point in limiting everyone else by preserving old packages for timeless information.

In the end, the debate about the future of books is really about the relative performance of competing technologies. Books are designed by people, as are computers. There are plenty of examples of apparently irreconcilable disagreements over a new technology disappearing once the technology catches up to the specifications of what it's replacing. Many people used to religiously proclaim that they couldn't edit text on a computer screen; once the resolution of a screen began to approach that of the eye at a comfortable reading distance, the discussion began to go away. Now writing a text by hand is the exception rather than the rule. Such passionate debates get settled not by persuasion, but by technical progress making them increasingly irrelevant.

Along the way, the presumptions of a new technology must usually be tempered by the wisdom embodied in an old one. In the early days of the internal combustion engine it was interesting to race horses and cars. Now we have supersonic cars, but no one is arguing for the abolition of horses. Although horses are no longer the fastest means of transportation, no current car can recognize its owner with a glance, or choose a path through a narrow mountain pass, or be left in a meadow to refuel itself, or make a copy of itself when it begins to wear out. Cars still have a long way to go to catch up to horses.

The computers that are meant to replace books are destined to be transformed even more by the books. Reading is too varied and personal to be bounded by the distinction between digital and analog. I have some beloved books that I'll never want to browse through in any other form, and piles of books threatening to topple over onto my desk that I'd love to access electronically, and some inscrutable texts that I'll never get anywhere with until they're available with active annotations and on-line connections.

Choosing between books and computers makes as much sense as choosing between breathing and eating. Books do a magnificent job of conveying static information; computers let information change. We're just now learning how to use a lot of new technology to match the performance of the mature technology in books, transcending its inherent limits without sacrificing its best features. The bits and the atoms belong together. The story of the book is not coming to an end; it's really just beginning.

Digital Expression

A few years ago I found myself on the stage of one of Tokyo's grandest concert halls. I wasn't going to perform; Yo-Yo Ma was, if I could fix his cello bow in time. The position sensor that had worked so well at MIT was no longer functioning after a trip around the world.

To a casual observer, Yo-Yo's instrument looked like a normal cello that had lost a few pounds. The one obvious clue that something was out of the ordinary was the racks of computers and electronics behind it. These made the sound that the audience heard; the cello was really just an elaborate input device.

Sensors measured everything Yo-Yo did as he played. A thin sandwich of flexible foam near the end of the bow detected the force being applied to it, and a rotary sensor measured the angle of his wrist holding the bow. Strips under the strings recorded where they contacted the fingerboard. Polymer films measured how the bridge and top plate of the instrument vibrated. In a conventional cello these vibrations get acoustically amplified by the resonant cello body that acts like a loudspeaker; here they were electronically amplified by circuits in the solid body. The cello itself made no audible sound. A small antenna on the bridge sent a radio signal that was detected by a conducting plastic strip on the bow, determining the position of the bow along the stroke and its distance from the bridge.

Earlier that day we had spent a few hours unpacking all of the shipping crates, attaching the many cables, and booting up the computers. Everything was going fine until we tried the bow. The position sensor that I had developed to measure its position was erroneously indicating that the bow was fluctuating wildly, even though it wasn't actually moving. Fortunately, I had packed a small laboratory of test and repair equipment. Given the amount of new technology being used in a hostile environment (a concert hall stage), something unexpected was certain to happen. Probing the signals let me narrow the problem down to the fine cable coming from the bow antenna. Opening up its protective insulation and shielding exposed an erratic connection, which I could patch around with a bit more surgery on the inner conductor. After I packed everything back up, the signals checked out okay, and I could resume breathing.

There was an expectant buzz in the full hall as we connected the computers, the cello, and Yo-Yo. The second unexpected event of the evening came as I began to make my way back to the control console where we were going to do some final system checks. I was only a few feet into the audience when I heard the unmistakable sound of Yo-Yo Ma controlling racks of electronics. The concert had begun. As some newfound friends made room to squeeze me in, I felt the way I imagined I will feel when my children leave home. The instrument had become his more than ours; he no longer needed us. As he played, the hall filled with familiar cello lines and unfamiliar sonic textures, with recognizable notes and with thick layers of interwoven phrases, all controlled by him. I sat back to enjoy the performance as a listener, no longer an inventor.

I was equally surprised and delighted to be there. This project was both an artistic and a technical experiment, asking how the old technology of a cello could be enhanced by the new technology of a computer. Yo-Yo's real allegiance is to communicating musical expression to a listener; the cello is just the best way he's found to do that. By introducing intelligence into the instrument we could let him control more sounds in more ways without sacrificing the centuries of experience reflected in its interface and his technique. This was an ideal way to explore the role of technology in music: there was an enormous incentive to succeed, and it would be easy to tell if we didn't.

My contribution had begun years earlier with my love/hate relationship with bassoon reeds. As an enthusiastic amateur musician I was just good enough to be able to get hopelessly out of my depth. One of my perennial stumbling blocks was reed making. To create a reed, cane from a particular region in France is seasoned, shaped, cut, trimmed, wrapped, sealed, shaved, tested, aged, played, shaved some more, played again, trimmed a bit, on and on, until, if the reed gods smile on you, the reed finally speaks over the whole range of pitch and volume of the bassoon. Almost imperceptible changes in the thickness of the reed in particular places can have a salutary or devastating effect on the color of the sound, or the dynamics of the attacks, or the stability of the pitch, or the effort required to make a sound. The reward for all of this trouble is playing one of the most expressive of all musical instruments. The same sensitivity that makes it so difficult to put together a good reed also lets the reed respond to subtle changes in the distribution of pressure and airflow as it is played. When it works well I have trouble telling where I stop and the instrument begins.

For a few weeks, that is, until the reed becomes soggy and the whole process starts all over again. The cane softens, cracks, and wears away, reducing the reed to buzzing noodles. This is why I was ready to embrace the convenience of electronic music synthesizers when I first encountered them. I was a graduate student at Cornell working on experiments in the basement of the Physics building, crossing the street to go to the Music building to play my bassoon for therapy. I didn't realize that I was literally following in the footsteps of Bob Moog.

A few decades earlier he had walked the same route, leaving his experiments in the Physics building to invent the modern synthesizer. He was struck by a parallel between the human ear and the transistor, which was just becoming commercially available. Both devices respond exponentially rather than linearly. This means that each time the amplitude or frequency of a sound is doubled, the perceived intensity or pitch increases by the same fixed amount. A low octave on a piano might span 100 to 200 Hz; another set of twelve keys farther up the keyboard might go from 1000 to 2000 Hz. The same thing is true of the current flowing through a transistor in response to an applied control voltage. This is what let Bob make circuits that could be played like any other musical instrument.

Moog began building modules with transistor circuits to control the frequency of oscillators, the volume of amplifiers, or the cutoff of filters. Patching together these modules resulted in great sounds that could be used in musically satisfying ways. As Bob's musical circuits grew more and more capable, he spent less and less time in the physics lab, until his advisor got rid of him by giving him a degree. This was fortunate for the rest of the world, which was transformed by the company Bob then founded. The progeny of the Moog synthesizer can be heard in the electronic sounds in almost any popular recording.

At Cornell this legacy came to life for me when I encountered one of Bob's early synths still in use in the basement of the Music building. It belonged to David Borden, the composer who had used

Bob's original synthesizers to start the first electronic music ensemble. David also had a useful talent for breaking anything electronic by looking at it; when one of Bob's modules could survive the Borden test, then it was ready to be sold.

David was in the process of updating Cornell's studio to catch up to the digital world. The studio had languished since Bob's time as the department specialized in the performance practice of early acoustic musical instruments, leaving it with the earliest electronic ones also. As David filled the studio with the latest computerized synths, I found it liberating to come in and push a button and have the most amazing sounds emerge. Even better, the sound was there the next day, and the next month; I didn't have to worry about it becoming soggy.

I quickly realized that was also the problem. The nuance that made my bassoon so beloved was lost in the clinical perfection of the electronic sounds. Unlike the bassoon, there was no ambiguity as to what part I did and what part it did: I pressed the button, it beeped. I began to give up on the technology and go back to my real instrument.

As I walked back and forth between the Music building and the Physics building I realized that I was doing something very silly. A bassoon is governed by the same laws of physics as the rest of the universe, including my physics experiments. It may be a mystery why we like to listen to a bassoon, but the equations that govern its behavior are not. Perhaps I could come up with a set of measurements and models that could reproduce its response, retaining what I like about the bassoon but freeing me from the Sisyphian task of reed making.

Directly comparing a bassoon and an electronic synthesizer makes clear just how impressive the bassoon really is. Even though its design is centuries old, it responds more quickly and in more complex ways to finer changes in more aspects of the musician's actions. It's not surprising that when playing a synthesizer one feels that something's missing.

From across the street in the Physics department, it occurred to me that, instead of leaving the lab to make music, I could stay there and use its resources to update the synthesizer with everything that's been learned about sensing and computing since Bob Moog's day. As I began the preliminary analysis that precedes any physics experiment, I was startled to realize that for the purpose of making music, computers are beginning to exceed the performance of nature.

There are relevant limits on both the instrument and the player. Displacements of a cello bow that travel less than about a millimeter or happen faster than about a millisecond are not audible. This means that the data rate needed to determine the bow's position matches that generated by a computer mouse. Even including the extra information needed to describe the bow pressure and angle, and the positions of the fingers on the strings, the amount of gestural input data generated by a player is easily handled by a PC. If we use the specifications of a CD player to estimate the data rate for the resulting sound, this is also easily handled by a PC. Finally, the effective computational speed of the cello can be found by recognizing that the stiffness and damping in the materials that it is made of restrict its response to vibrations that occur over distances on the order of millimeters, at a frequency of tens of thousands of cycles per second. A mathematical model need not keep track of anything happening smaller or faster than that. Dividing the size of the cello by these numbers results in an upper limit of billions of mathematical operations per second for a computer to model a cello in real time. This is the speed of today's supercomputers, and will soon be that of a fast workstation.

The conclusion is that with appropriate sensors, a computer should be able to compete with a Stradivarius. By the time I appreciated this I was a Junior Fellow of the Harvard Society of Fellows. The regular formal dinners of the Society provided an ideal setting to develop this argument, but not a Stradivarius to try it out on. That came when I met a former Junior Fellow, Marvin Minsky, at one of the dinners. He called my bluff and invited me down to the other end of Cambridge to visit the Media Lab and take the experiment seriously.

Marvin introduced me to Tod Machover, a composer at the Media Lab. Starting with his days directing research at IRCAM, the pioneering music laboratory in Paris, Tod has spent many years designing and writing for smart instruments. Classically, music has had a clear division of labor. The composer puts notes on a page, the musician interprets the shorthand representation of the composer's intent by suitable gestures, and the instrument turns those gestures into sounds. There's nothing particularly fundamental about that arrangement. It reflects what has been the prevailing technology for distributing music—a piece of paper. Notes are a very low level for describing music. Just as computer programmers have moved from specifying the primitive instructions understood by a processor to writing in higher-level languages matched to application domains such as mathematics or bookkeeping, a more intelligent musical instrument could let a composer specify how sounds result from the player's actions in more abstract ways than merely listing notes. The point is not to eliminate the player, it is to free the player to concentrate on the music.

Tod had been discussing this idea with Yo-Yo, who was all too aware of the limits of his cello. While he's never found anything better for musical expression, his cello can play just one or two notes at a time, it's hard to move quickly between notes played at opposite ends of a string, and the timbral range is limited to the sounds that a bowed string can make. Within these limits a cello is wonderfully lyrical; Yo-Yo was wondering what might lie beyond them.

As Tod, Yo-Yo, and I compared notes, we realized that we were all asking the same question from different directions: how can new digital technology build on the old mechanical technology of musical instruments without sacrificing what we all loved about traditional instruments? We decided to create a new kind of cello, to play a new kind of music.

My job was to find ways to measure everything that Yo-Yo did when he played, without interfering with his performance. Sensors determined where the bow was, how he held it, and where his fingers were on the strings. These were connected to a computer programmed by a team of Tod's students, led by Joe Chung, to perform low-level calibration (where is the bow?), mid-level analysis (what kind of bow technique is being used?), and high-level mapping (what kind of sound should be associated with that action?). The sounds were then produced by a collection of synthesizers and signal processors under control of the computer.

Tod, who is also a cellist, wrote a piece called *Begin Again Again* . . . that looked like conventionally notated cello music (since that is what Yo-Yo plays), but that also specified the rules for how the computer generated sounds in response to what Yo-Yo did. Yo-Yo's essential role in the development was to bridge between the technology and the art. He helped me understand what parts of his technique were relevant to measure, and what parts were irrelevant or would be intrusive to include, and he helped Tod create musical mappings that could use these data in ways that were artistically meaningful and that built on his technique.

While the physical interface of the cello never changed, in each section of the resulting piece the instructions the computer followed to make sounds did. In effect, Yo-Yo played a new instrument in each section, always starting from traditional practice but adding new capabilities. For example, an important part of a cellist's technique is associated with the bow placement. Bowing near the bridge (called *ponticello*) makes a bright, harsh sound; bowing near the fingerboard makes a softer, sweeter sound. In one section of the piece these mappings were extended so that playing *ponticello* made still brighter sounds than a cello could ever reach. Another essential part of playing a cello is the trajectory of the bow before a note starts and after it ends. Not unlike a golf swing or baseball throw, the preparation and follow-through are necessary parts of the bow stroke. In addition, these gestures serve as cues that help players communicate with each other visually as well as aurally. These influences came together in a section of the piece that used the trajectory of his bow to launch short musical phrases. The location and velocity of the bow controlled the volume and tempo of a sequence of notes rather than an individual note. Yo-Yo described this as feeling exactly like ensemble playing, except that he was the ensemble. The computer would pick up on his bowing cues and respond appropriately, in turn influencing his playing.

Developing the instrument was a humbling experience. I expected to be able to use a few off-the-shelf sensors to make the measurements; what I found was that the cello is such a mature, tightly integrated system that most anything I tried was sure to either fail or make something else worse. Take a task as apparently simple as detecting the motion of the strings. The first thing that I tried was inspired by the pickup of an electric guitar, which uses a permanent magnet to induce a current in a moving string, in turn creating a magnetic field that is detected by a coil. Since a cello string is much farther from the fingerboard than a guitar string, I had to design a pickup with a much stronger magnet and more sensitive amplifier. This let me make the measurement out to the required distance, obtaining a nice electrical signal when I plucked the string. When this worked, I proudly called the others in to try out my new device.

To my horror there was no signal when the string was bowed, even though I had just seen it working. Some more testing revealed the problem: bowing made the string move from side to side, but my sensor responded only to the up-and-down motion of the string that resulted when I plucked it.

Plan B was to place between the strings and the bridge a thin polymer sheet that creates a voltage when compressed. By carefully lining the bridge with the polymer I was able to get a nice measurement of the string vibration. Pleased with this result, and ignoring recent experience, I called everyone back in to see it work. And then I watched them leave again when the signal disappeared as the cello was bowed. Here the problem was that the bowing excited a rocking mode of the bridge that eliminated most of the constraint force that the polymer strips were measuring.

A similar set of difficulties came up in following the movement of the bow. Sonar in air, sonar in the strings, optical tracking, radars, each ran afoul of some kind of subtle interaction in the cello. It was only after failing with these increasingly complicated solutions that I found the simple trick we used. I was inspired by a baton for conducting a computer that was developed by Max Mathews, a scientist from Bell Labs who was the first person to use a computer to make music. His system measured the position of the baton by the variation in a radio signal picked up in a large flat antenna shaped like a backgammon board. This was much too big to fit on a cello, but I realized that I could obtain the same kind of response from a thin strip of a material placed on the bow that was made out of poor conductor so that the signal strength it received from a small antenna on the bridge would vary as a function of the bow position.

The most interesting part of the preparations were the rehearsals at MIT where the emerging hardware and software came together with the composition and the musical mappings. Each element evolved around the constraints of the others to grow into an integrated instrument. When everything was working reasonably reliably, we packed up and headed off for the premiere performance at Tanglewood in western Massachusetts. It was hard to miss our arrival when, for the first time, Yo-Yo went on stage to try out the whole system.

A cello can only sound so loud. Beyond that, players spend a lifetime learning tricks in how they articulate notes to seem to sound louder and to better cut through an accompanying orchestra. There's nothing profound about this aspect of technique; it's simply what must be done to be heard with a conventional cello. But we had given Yo-Yo an unconventional cello that could play as loud as he wanted, and play he did. He unleashed such a torrent of sound that it threatened to drown out the concert going on in the main shed. This was great fun for everyone involved (except perhaps the people attending the concert next door).

A similar thing happened the first time he tried the wireless bow sensor. Instead of playing what he was supposed to be rehearsing, he went off on a tangent, making sounds by conducting with his bow instead of touching the strings at all. After a lifetime of thinking about the implications of how he moved his bow through the air, he could finally hear it explicitly.

As the rehearsals progressed we found that eliminating many other former constraints on cello practice became easy matters of system design. It's taken for granted that the cellist, conductor, and audience should all hear the same thing, even though they're listening for different things, and are in different acoustic environments. We found that Yo-Yo was best able to play with a sound mix in his monitor speakers that highlighted his performance cues, different from the sound mix that filled the hall. Such a split is impossible to do with an acoustic cello but was trivial with ours.

The biggest surprise for me was the strength of the critical reaction, both positive and negative, to connecting a computer and a cello. Great musicians loved what we were doing because they care about the music, not the technology. They have an unsentimental understanding of the strengths and weaknesses of their instruments, and are unwilling to use anything inferior but are equally eager to transcend all-too-familiar limitations. Beginners loved what we were doing because they don't care about the technology either, they just want to make music. They're open to anything that helps them do that. And in between were the people who hated the project.

They complained that technology was already intruding on too many parts of life, and here we were ruining one of the few remaining unspoiled creative domains. This reaction is based on the curious belief that a computer is technology, and a cello isn't. In fact, musical instruments have always been improved by drawing on the newest available technologies. The volume of early pianos was limited by the energy that could be stored in the strings, which in turn was limited by the strength of the wood and metal that held them. Improvements in metallurgy made possible the casting of iron frames that could withstand tons of force, creating the modern piano, which is what enabled Beethoven to write his thunderous concertos. Popular musical instruments have continued to grow and change, but classical instruments have become frozen along with much of their repertoire. The only role for new technology is in helping people passively listen to a small group of active performers.

What our critics were really complaining about were the excesses of blindly introducing inferior new technology into successful mature designs. Given that a Stradivarius effectively computes as fast as the largest supercomputer, it's no wonder that most computer music to date has lacked the nuance of a Strad. Just as with electronic books, rather than reject new technology out of hand it is much more challenging and interesting to ask that it work better than what it intends to replace.

Here our report card is mixed. We made a cello that let Yo-Yo do new things, but our instrument couldn't match his Strad at what it does best. I view that as a temporary lapse; I'll be disappointed if we can't make a digital Stradivarius in the next few years. That's admittedly a presumptuous expectation, needing an explanation of how we're going to do it, and why.

Luthiers have spent centuries failing to make instruments that can match a Strad. Their frustrating inability to reproduce the lost magic of the Cremona school has led them to chase down many blind alleys trying to copy past practice. The nadir might have been an attempt to use a rumored magic ingredient in the original varnish, pig's urine. Better results have come from trying to copy the mechanical function of a Strad instead of its exact design.

By bouncing a laser beam off of a violin as it is played, it is possible to measure tiny motions in the body. These studies have led to the unexpected discovery that modes in which the left and right sides of the instrument vibrate together sound bad, and modes in which they move asymmetrically sound good. This helps explain the presence of the tuning peg inside the instrument, as well as why the two sides of a violin are not shaved identically. Instruments built with frequent testing by such modern analytical tools have been steadily improving, not yet surpassing the great old instruments but getting closer.

Progress in duplicating a Stradivarius is now coming from an unexpected quarter. Physicists have spent millennia studying musical instruments, solely for the pleasure of understanding how they work. From Pythagoras's analysis of a vibrating string, to Helmholtz's discovery in the last century of the characteristic motion of a string driven by the sticking and slipping of a bow (found with a vibrating microscope he invented by mounting an eyepiece on a tuning fork), to Nobel laureate C. V. Raman's more recent study of the role of stiffness in the strings of Indian musical instruments, these efforts have helped apply and develop new physical theories without any expectation of practical applications. The work generally has been done on the side by people who kept their day jobs; a recent definitive text on the physics of musical instruments starts with a somewhat defiant justification of such an idle fancy. Computation is now turning this study on its head.

The world isn't getting any faster (even though it may feel that way), but computers are. Once a computer can solve in real time the equations describing the motion of a violin, the model can replace the violin. Given a fast computer and good sensors, all of the accumulated descriptions of the physics of musical instruments become playable instruments themselves. This is why I expect to be able to create a digital Stradivarius. I'm not smarter than the people who tried before and failed; I'm just asking the question at an opportune time.

I also think that I know the secret of a Strad. When we sent the data from Yo-Yo's sensors to almost any sound source, the result still sounded very much like Yo-Yo playing. The essence of his artistry lies in how he shapes the ten or so attributes of a note that are available to him, not in the details of the waveform of the note. Much of the value of a Strad lies not in a mysterious attribute of the sound it makes, but rather in its performance as a controller. Where a lesser instrument might drop out as a note is released or waver when a note is sustained, a Strad lets a skilled player make sharp or soft attacks, smooth or abrupt releases. This translates into effective specifications for the resolution and response rate of the interface, something we've already found that we can match. This is the second reason that I'm optimistic about making a digital Strad: much of the problem lies in tractable engineering questions about sensor performance.

For a computer to emulate a Stradivarius it must also store a description of it. Right now this sort of problem is solved most commonly in electronic musical instruments by recording samples of sounds and playing them back. This is how most digital pianos work. The advantage is that a short segment of sound is guaranteed to sound good, since it's a direct replica of the original. The disadvantage is readily apparent in listening to more than a short segment. The samples can't change, and so they can't properly respond to the player. The result lacks the fluid expression of a traditional instrument.

An alternative to storing samples is to use a computer to directly solve a mathematical model of the instrument. This is now possible on the fastest supercomputers and will become feasible on more widely accessible machines. Aside from the demand for very powerful computers, the difficulty with this approach is that it's not much easier than building an instrument the old-fashioned way. Even if the wood and strings are specified as software models they must still match the properties of the real wood and strings. That's exactly the problem that's remained unsolved over the last few centuries.

A better alternative lies between the extremes of playing back samples and solving physical models. Just as luthiers began to progress when they moved their focus from how a Stradivarius is built to how it performs, the same lesson applies to mathematical modeling. Instead of trying to copy the construction of an instrument, it's possible to copy its function. We can put our sensors on a great instrument and record the player's actions along with the resulting sound. After accumulating such data covering the range of what the instrument is capable of, we can apply modern data analysis techniques to come up with an efficient model that can reproduce how the instrument turns gestures into sounds. At that point we can throw away the instrument (or lovingly display it to admire the craftsmanship and history) and keep the functionally equivalent model. You can view this as sampling the physics of the instrument instead of sampling the sound. Once the model has been found, it can be played with the original sensors, or used in entirely new ways. Soon after we were first able to model violin bowing with this kind of analysis I was surprised to come into my lab and see my student Bernd Schoner playing his arm. He had put the bow position sensor into the sleeve of his shirt so that he could play a violin without needing to hold a violin.

Granting then that a digital Stradivarius may be possible in the not-too-distant future, it's still fair to ask what the point is. The obvious reason for the effort is to make the joy of playing a Strad accessible to many more people. There are very few Strads around, and they're regularly played by even fewer people. Even worse, the instruments that aren't in routine use because they're so valuable suffer from problems similar to those of an automobile that isn't driven regularly. If we can match the performance of a Strad with easily duplicated sensors and software, then playing a great instrument doesn't need to be restricted to a select few.

Beyond helping more people do what a small group can do now, making a digital Stradivarius helps great players do things that no one can do now. As we found in the project with Yo-Yo, traditional instruments have many limitations that just reflect constraints of their old technology. By making a virtual Stradivarius that is as good as the real thing, it's then possible to free players from those constraints without asking them to compromise their existing technique.

Making a Strad cheaper, or better, is a worthy goal, but one that directly touches only the small fraction of the population that can play the cello. There's a much more significant implication for everyone else. Right now you can listen to a recording of Bach's cello suites, or you can play them yourself. It takes a lifetime to learn to do the latter well, and along the way the suites don't sound particularly good. But when a Stradivarius merges with a PC then this divide becomes a continuum. The computer could emulate your CD player and play the suites for you, or your bowing could control just the tempo, or you could take over control of the phrasing, on up to playing them the old-fashioned way.

A wealthy executive was once given the chance to conduct the New York Philharmonic; afterward when asked how he did, a player commented, "It was fine; he pretended to conduct, and we pretended to follow him." That's exactly what a smart musical instrument should do: give you as much control as you want and can use, and intelligently fill in the rest.

Now this is an exciting consequence of bringing computing and music together. It used to be that many people played music, because that was the only way to hear it. When mass media came along, society split into a small number of people paid to be artistically creative and a much larger number that passively consumes their output. Reducing the effort to learn to play an instrument, and opening up the space between a PC, a CD player, and a Stradivarius, points to the possibility that far more people will be able to creatively express themselves. Improving the technology for making music can help engage instead of insulate people.

There's a final reason why it's worth trying to make a digital Stradivarius: it's hard. The most serious criticism of so many demos of new interfaces between people and computers—say, the latest and greatest virtual reality environment—is that it's hard even to say if they're particularly good or bad. So what if they let you stack some virtual blocks? The situation is very different with a cello, where the centuries of wisdom embodied in the instrument and the player make it very easy to tell when you fail. The discipline of making an instrument good enough for Yo-Yo Ma ended up teaching us many lessons about the design of sensors and software that are much more broadly applicable to any interaction with a computer.

Constraints really are key. If you venture off without them it's easy to get lost. One of the most depressing days I've ever spent was at a computer music conference full of bad engineering and bad music. Too many people there told the musicians that they were engineers, and the engineers that they were musicians, while doing a poor job at being either. Increasingly, we can make sensors for, and models of, almost anything. If this kind of interactivity is done blindly, everything will turn out sounding brown. Technologists should be the last people guiding and shaping the applications; accepting the challenge presented by a Stradivarius is one demanding way to ground the effort.

Toward the end of the project with Yo-Yo, I asked him when he would be ready to leave his cello behind and use our instrument instead. His answer was instant. Rather than speculate about the irreproducible joys of his Strad, he jumped to the practical reality of making music. He tours constantly and needs to be able to get out of a plane, open his case, and start playing. Our system took the form of several cases of hardware tended to by a small army of students, and it took a few hours to boot up and get working. It's not until our technology can be as unobtrusive and invisible as that of the Strad that he could prefer it instead. It must always be available for use, never fail, and need no special configuration or maintenance. That imperative to make technology so good that it disappears lies behind much of the work in this book.

Wear Ware Where?

Steve Mann was one of the Media Lab's best-dressed students. It's easy to recognize Steve because you don't see him; you see his computer. His eyes are hidden behind a visor containing small display screens. He looks out through a pair of cameras, which are connected to his displays through a fanny pack full of electronics strapped around his waist. Most of the time the cameras are mounted in front of his eye, but when he rides a bicycle he finds that it's helpful to have one of his electronic eyes looking backward so he can see approaching traffic, or in a crowd he likes to move an eye down to his feet to help him see where he's walking.

Because the path an image takes to reach his eyes passes through a computer, he's able to transform what he sees. He can crank up the brightness to help him see in the dark; some days he likes to see like a bug and have the background fade out so that he perceives only moving objects; and over time he's found that he prefers to have the world appear rotated on its side. The reality that he perceives becomes something that can be adjusted under software control; once he's had enough of seeing like a bug, he can try out any other scheme for viewing the world that he can think of.

The rest of Steve's body is also wired. A small handheld keyboard lets him read and write without looking away from whatever else he is doing. It has one key for each finger; by pressing the keys down in combinations he is able to type as fast as he could on a conventional keyboard. Beyond his hands, his clothing is full of sensors connected to other parts of his body. Measurements of his heart rate, perspiration, and footsteps let him add a dashboard for his body to the display in his glasses, warning him when he's running hot and needs to relax.

Since a computer displays and records Steve's world, he can enhance more than just his senses. His memory is augmented by the accumulated data in his system. By storing and analyzing what he writes, and sees, and hears, the computer can help him recall when he last met someone, what they spoke about, and what he knows about him or her. Or, if he's fixing one of his ham radios, he can see a circuit diagram at the same time he sees a circuit board.

Steve is not an island; an antenna on his head links him into the network. His body is both a licensed video transmitter and an Internet node. This lets him send and receive information, and access faster computers and bigger databases than he can carry around. Not only can he exchange e-mail with friends during the course of a day, he can exchange his senses. The data that goes to his display screens can just as easily be sent around the world; he maintains a Web page with images he's recently seen. Steve's wife can look out through his eyes when he's shopping and help him select fruit, something he doesn't do well. If he's walking down a dark street late at night, he can have some friends look over his virtual shoulder to help keep a remote eye on him.

Working in the Media Lab over the last few years has been like living through a B movie, *The Invasion of the Cyborgs*. One by one our conventionally equipped students have been disappearing, replaced by a new generation that, like Steve, is augmented by wearable computers. At first I found it disconcerting to teach to people who were seeing things that I wasn't, but I soon found that the enhancements they were applying to me helped me better communicate with them. When one of Steve's colleagues, Thad Starner, took a seminar from me, he insisted that I make my lecture notes available on the network for him. He wore glasses that projected a display that allowed him to see my text and me at the same time, so that he took notes on a copy of the lecture that appeared to him to be floating next to me. Because of the apparently intrusive technology intervening between us I actually had more of his attention than the students flipping through paper copies.

When Thad, Steve, and their peers started wearing computers, people would cross the street to get away from them. Now people cross the street to get to them, to find out where they can buy one for themselves. Wearable computers are a revolution that I'm certain will happen, because it already *is* happening. Three forces are driving this transition: people's desire to augment their innate capabilities, emerging technological insight into how to embed computing into clothing, and industrial demand to move information away from where the computers are and to where the people are.

FedEx runs one of the world's largest airlines. Unlike an ordinary airline that usually gets a few weeks' notice between the time a ticket is sold and a flight departs, FedEx has just a few hours after it receives a package to make sure that planes are in the right place, at the right time, with room for the package. Each step in this nocturnal logistical miracle is invariably a race against time, as packages are sorted and packed into planes preparing to depart. Literally seconds count, as delays at any point can ripple through the system. Ideally, the planning of the fleet's routes and manifests should begin the instant that a driver picks up a package. To be useful, this must be done without encumbering the drivers, who need to be able to speed through crowded offices with awkward loads.

This is why FedEx has been a pioneer in the industrial application of wearable computers. Their package scanners and data networks are moving from the central hubs, to the trucks, to the drivers, so that as soon as an item is collected its data can be racing ahead of it on a wireless link to prepare its itinerary.

Maintaining a fleet of airplanes like FedEx's presents another industrial demand for wearable computers. The owner's manual for a 747 weighs a few tons, and because of the need to control access for safety and liability, it can't even be kept in the hangar where it is needed. It certainly doesn't fit on a plane that must maximize revenue per pound. Keeping the manual up-to-date is a full-time job, with life-and-death implications that necessitate recording a careful paper trail of changes. Complex maintenance procedures need to be followed by mechanics working in confined spaces, once again keeping a record of each operation they perform. For these reasons aviation mechanics are beginning to wear computers with display glasses that let them follow instructions without looking away from their work, disk drives that can store a mountain of manuals (one CD-ROM holds about a million pages of text), and network connections that automate the record keeping.

While an airplane mechanic is willing to strap on pounds of computer gear in order to get access to necessary information, and a Media Lab grad student will do the same as a lifestyle choice, most everyone else is unlikely to be willing to walk around looking like they were wearing flypaper when a hurricane hit an electronics store. Fortunately, we're learning how to provide the functionality of a wearable computer without the inconvenience of the crude early models.

It's a safe assumption that the clothes you wear are near a body (yours). One of my former grad students, Tom Zimmerman (now at IBM), realized that this trivial observation has a remarkable implication. Tom had come back to grad school after inventing the

Data Glove, a glove full of wires that measures the location and configuration of a hand. The company he cofounded, VPL, developed this glove into the interface that helped launch the concept of virtual reality, letting people use physical interactions to manipulate virtual worlds. But Tom was dissatisfied with the cumbersome wires and wanted to find new ways to build interfaces that are as natural as the physical world.

Tom arrived at the Media Lab around the time that we were working on a collaboration with Tod Machover and the violinist Ani Kavafian. After the project with Yo-Yo I thought that I understood everything there was to know about measuring the position of the bow of a stringed instrument. Moments after putting the sensors on Ani's violin it was clear that something was very wrong: we were finding the location of her hand, not the bow. Yo-Yo plays his Strad sitting down, with his hand above the bridge; Ani plays hers standing up, with her hand beyond the bridge. Something about this change in geometry was leading to an interaction between her hand and the electric field that was supposed to detect the bow.

I was puzzled by this artifact because the signal went in the wrong direction. I expected that by placing her hand between the transmitting and receiving electrodes she would improve the coupling and make the signal stronger, but her hand had exactly the opposite effect and made the signal weaker. I tried unsuccessfully to solve the complex equations that are needed to predict this behavior; then I did something very smart: I went on a trip to Europe and left Tom alone in the lab. While I was away, he walked down to the corner deli and had them fill a glove with hamburger. While they were left wondering about Tom's odd taste, he brought this armless hand back to the lab and confirmed that it caused the signals to move in the direction that I had expected—up. When he connected the hand to a wire arm, the interaction then flipped to make the signal go down. The source of our problem was immediately clear:

part of Ani's body was in the field and part was out; she was guiding the field away from the receiver and out to the room.

Tom then realized that we should be able to detect the part of the field that was passing through her body. This creates a tiny current (about 0.0000001 percent of what illuminates a lightbulb). To find such a small signal buried in all of the other possible sources of electrical interference, we used a circuit that was matched to the pattern that was being transmitted. Therefore, if we could find that pattern, we could find other ones. In other words, we could transmit data through a body. The bug could become quite a feature.

We soon had a prototype working. A transmitting unit, about the size of a deck of cards, had a pair of electrodes on opposite sides that created an electric field. Placed near a body, for example in a shoe, the field changed the body's average voltage by a tiny amount. A similar receiving unit could measure this voltage change at another part of the body. By varying the tiny voltage, a message could be exchanged. This means that a computer in a shoe could send data to a display in a wristwatch without needing wires. There are Wide Area Networks (WANs) to link up cities, and Local Area Networks (LANs) to link up buildings; we had created a Personal Area Network (PAN) to connect parts of a body.

Unlike a conventional wireless radio, PAN keeps personal data in the body, where it belongs. The radio spectrum is one of the scarcest of modern resources, as more and more applications vie for the available channels. One way to use the spectrum efficiently is to divide it up geographically, such as the roughly one-kilometer cells used for mobile telephones. PAN shrinks the cells down to one body. A roomful of people could be using it without interfering with each other, or worrying about someone surreptitiously eavesdropping—unless they approach close enough to pick the bits out of their pockets.

Within one body, PAN provides a means to get rid of the wires in a wearable computer. Between bodies, it can do much more. Two people using it could exchange an electronic business card just by shaking hands. A hand on a doorknob could send the data to unlock a door, or the act of picking up an airport telephone could download the day's messages. The contents of a package could be read automatically as it is picked up.

Right now, gestures are either physical (exchanging a printed card) or logical (exchanging e-mail). With PAN, physical gestures can take on logical meaning. We've spent millennia as a species learning how to manage our physical interactions with people and things. We have a comfortable protocol for when and where to shake hands. Rather than decouple such familiar gestures from the digital world, PAN helps create a reality that merges the logical and physical components. The World Wide Web helps you to think globally but not to act locally; a Person Wide Web is needed for that.

The biggest surprise about PAN is that it has been such a surprise. There's very little precedent for this idea, which is so obvious in retrospect. In the postlude to *3001: The Final Odyssey,* Arthur C. Clarke complains that he thought he had invented the idea of exchanging data through a handshake as a capability for the next millennium, and was mortified to later find out that we were already doing it.

Dick Tracy might have had his wristwatch TV, but the real implications of wearing computers have been too unexpected to be anticipated by fiction. Once a computer becomes continuously available it literally becomes part of the fabric of our lives rather than just an appliance for work or play. It is a very different conception of computation.

The closest fictional predecessor to PAN was Maxwell Smart's shoe phone, which he had to take off (invariably at the most inopportune times) if he wanted to call the Chief. PAN now lets Max keep his shoes on, and shoes are in fact an ideal platform for computing. You almost always have your shoes with you; you don't need to remember to carry them along when you leave the house in the morning. There's plenty of space available in a shoe for circuitry—no companies are yet fighting over access to your shoes, but they will. When you walk, watts of power pass through your feet and get dissipated in abrading your shoes and impacting the ground; it's possible to put materials in a shoe that produce a voltage when they are flexed, which can be used as a power source. With realistic generation efficiencies of a few percent there's plenty of energy available for a low-power computer. Instead of lugging around a laptop's power supply and adapters, and feeding it batteries, you just need to go for a walk every now and then, and feed yourself. Finally, PAN lets a shoe computer energize accessories around your body, communicating with a wristwatch or eyeglasses. Running down this list of specifications, it's apparent that a foot bottom makes much more sense than a lap top. Shoe computers are no joke.

Wearable computers are really just a natural consequence of the personalization of computation. The original impersonal computers were mainframes, locked away in remote rooms. Then came minicomputers, which could be shared by one workgroup. From there came the PC, a computer used by a single person. This allowed much more personal expression, as long as you equate computing with sitting in front of a beige box with a cathode ray tube, keyboard, and mouse. Wearables finally let a computer come to its user rather than vice versa.

Joel Birnbaum, the thoughtful senior vice president for research and development at Hewlett Packard, notes that in the computer industry "peripherals are central." Most of the attention in the business, and press, has been focused on the components inside a computer (processor speed, memory size, . . .). But as PCs increasingly become commodities like toasters, which are more-or-less interchangeable and sold based on price, the emphasis is shifting to the external peripherals that people use to interact with the computer. From there it's a small step to focus on your accessories instead of the computer's: the eyeglasses that can serve as displays for your eyes, the earrings that can whisper a message into your ears. Right now the demand for such wearable components is being imperfectly met by home-brew parts and small start-up firms. Since high-tech industries abhor a product vacuum, this is sure to be filled by computer companies seeking to grow beyond their current saturated markets.

If computers have become like toasters, then software has become like toast. Interacting with a computer is all too often about as satisfying as interacting with a piece of toast, filling without being particularly stimulating, frequently overdone, and with a nasty habit of leaving messy crumbs of information everywhere. Just as wearable computers replace the homogeneity of uniform hardware with the ultimate in personalization, their intimate connection with their wearer helps do the same for software.

Computers currently have no idea whether you're happy or sad, tense or relaxed, bored or excited. These affective states have enormous implications for your relationship with the world; if you're in a good mood I'll tell you different things than if you're in a bad one. A person who cannot perceive affect is emotionally handicapped and cannot function normally in society, and the same is true of a computer.

Because they have no idea what state you're in, machines are uniformly cheerful, or bland, behavior that is just as annoying in a computer as it is in a person. Affect has very immediate implications. Studies have shown that if you're stressed, information delivered quickly helps you relax; if you're relaxed, information delivered quickly makes you stressed. Something as simple as the pace at which a computer delivers information should depend on your mood.

Steve Mann's thesis advisor Roz Picard has been finding that the kind of sensor data that a wearable can collect about its wearer (temperature, perspiration, muscle tension, and so forth), coupled with the techniques that have been developed in the last decade to let computers recognize patterns such as faces in images, enables a wearable computer to do a surprisingly good job of perceiving affect by analyzing the data that it records. But this shouldn't be very surprising: you routinely do the same with other people just by looking and listening.

As unsettling as it might be to contemplate a computer that is aware of your mood, the converse is to be forced to tolerate one that is sure to almost always respond to it inappropriately. Byron Reeves and Clifford Nass at Stanford have shown that most of what we know about the psychology of human interaction carries over to how people interact with computers; whatever our intentions, we already relate to computers as we do to people. The Stanford group has observed that most of the behavior that you can read about in psychology textbooks also occurs between people and computers: aggressive people prefer aggressive instructions from the machine while more withdrawn people prefer instructions that are phrased more gently, people are more critical in evaluating a computer if they fill out a survey on a second computer than if they fill it out on the one being evaluated, and so forth. Because we subconsciously expect computers to be able to understand us, we must give them the means to do so.

Even more disturbing than the prospect of a computer that knows how you feel is one that can share this most personal of information with other people, but here, too, the benefits can outweigh the risks. My colleague Mike Hawley worked with Harry Winston Jewelers to create what must be the world's most expensive computer display, a $500,000 diamond brooch with a heart-rate monitor that lets it pulse red in synchrony with its wearer's heartbeat. While you may not want to wear your heart on your sleeve, how about letting your jewelry tell a romantic interest when your heart skips a beat?

This fanciful brooch was part of the world's first wearable fashion show, held at the Media Lab in the fall of 1997. I began to realize that wearables were turning from geek-chic to high fashion when two of our grad students, Rehmi Post and Maggie Orth, figured out how to use conducting threads to embroider circuitry on fabric. Using filaments of Kevlar and stainless steel, they were able to program a computer-controlled sewing machine to stitch traces that could carry data and power, as well as sense the press of a finger. They call this *e-broidery*. It allowed them to get rid of the last of the apparent wires in a wearable, marrying the form and function into the garment. One of the first applications was a Levi's denim jacket that doubles as a musical instrument: a keypad sewn onto the front of the jacket lets the wearer call up rhythmic backing and play accompanying melodic fills. I suspect that Josh Strickon, the student who programmed the jacket's musical interface, is the first person from MIT to publish research in *Vogue*, appearing as a model to display the jacket.

Freed from the need to strap on external electronics, it was natural to start wondering what wearables could look like. This was a question for designers, not technologists, and so we turned to the fashion schools of the world. Guided by Thad's advisor Sandy Pentland, students from the Bunka Fashion College in Tokyo, Creapole in Paris, Domus in Milan, and the Parsons School of Design in New York submitted designs; the best were invited to come to the Media Lab to build them. Among the lively creations was a dress that played music based on the proximity of nearby bodies, dancing shoes full of sensors that controlled the sights and sounds of a performance, and a jacket made with an electronic ink that could change back and forth between a solid color and pinstripes under software (softwear?) control.

We weren't prepared to handle the enormous press interest in the event. Because clothing is so very personal, the reporting crossed over from the technical journalists usually seen around MIT to mainstream media. The many ensuing interviews conveyed a recurring sense of wonder, and fear, at the arrival of wearables. The reality of the show made clear that ordinary people may be wearing computers in the not-too-distant future, raising endless questions about the implications for how we're all going to live with them.

Privacy was a recurring concern. If it's already hard to sneeze without someone noticing it and selling your name in a database of cold-sufferers, what's going to happen if you carry a computer everywhere? Here I think that wearables are part of the solution, not the problem. Right now you're digitally unprotected; most of your communications and transactions are available to anyone who legally or illegally has the right equipment to detect them. From pulling charge slips out of a trash can to scanning cellular phone frequencies, it doesn't take much ambition to peer into people's private lives. Every time you call a toll-free number, or use a credit card, you're releasing quite a few bits of personal information to relative strangers.

There are cryptographic protocols that can ensure that you have complete control over who gets access to what information and when; the problem is using them. If you have to boot up your laptop each time you want to make a secure phone call, you're unlikely to use it. By wearing a computer that is routinely accessible, you can erect a kind of digital shield to help protect you by sending and receiving only what you intend. This won't be impermeable; privacy is not an absolute good, but a complex and personal trade-off. If you go into a store and make your electronic identity accessible, the store can give you better service because they learn about your preferences, and a discount because you've given them demographic information, but you've lost some privacy. If you turn your identity off, you can cryptographically perform a secure transaction that lets you make a purchase without the store learning anything about you, but it comes at a higher price and with worse service.

Another reason to consider sharing private information is for security. A cop walking the beat can see and hear only part of one street; wearables let many people see many places at the same time. While this certainly does raise questions about the spread of surveillance (students in the Media Lab made Steve Mann put a red light on his head to indicate when his eyes were on-line), a mugging is also quite an invasion of privacy. The balance between openness and privacy can't be settled in advance for all people in all situations; the technology needs to enable a continuous balance that can be adjusted as needed. On a lonely street late at night I want as many friends as possible to know what's happening around me; in a complex business negotiation I might want to make sure that some people elsewhere know everything and some know nothing; when I'm at home I may choose to let no one else have any access to me.

Along with loss of privacy, another fear is loss of autonomy. What happens if we come to rely on these machines? Thad, Steve, and their peers have confounded conventional wisdom about how long it's possible to be immersed in a computer environment. Early studies had suggested that after a few hours people start becoming disoriented and suffering ill effects; Thad and Steve have been wearing their systems all day long for years. What the studies missed was the extent to which their wearable computers have become an essential part of how these people function. Instead of performing an assigned task in a laboratory setting, they depend on their computers to see, and hear, and think. They are incomplete without them.

That's a less radical step than it might appear to be, because it's one we've all made. We're surrounded by technologies that we depend on to function in a modern society. Without the power for heat and light, the fabrics for protection from the environment, the printing for newspapers and books, the fuel for transportation, we could still survive, but it would be a much more difficult existence. It can be satisfying to throw off all of those things for a time and go camping in the woods, but even that is done by most people by relying on a great deal of portable technology for shelter and convenience. We are happy to consult a library to provide knowledge that we can't obtain by introspection; Steve and Thad go just a bit further to take advantage of a vast library of information that is continuously available to them.

Glasses are a familiar technology that we rely on to correct vision. Some people cannot function without the lenses that let them see near, or far, and some have vision problems that cannot be corrected by ordinary glasses. A wearable computer is equally useful for both, blurring the distinction between them. Whether or not you can read large or small print, you can't see in the dark, or see around the world, or look backward and forward at the same time. Wearables can certainly help with disabilities, but the range of what we consider to be abled and disabled is actually quite small compared to what both types of people can do by enhancing their senses with active processing. Many people are unwilling to give up these kinds of capabilities once they experience them. There's nothing new about that; the same is true of eyeglasses wearers.

A final concern with wearables is loss of community. A quick glance around an office where people who used to eat lunch together now sit alone surfing the Web makes clear that the expansion in global communication has come at the cost of local communication. The members of a wired family are already more likely than not to be found at home in separate rooms, typing away at their computers, even if they're just answering each other's e-mail. What's going to happen to interpersonal relationships if the computer becomes continuously accessible?

The responsibility for this loss of personal contact lies more with the computer on the desktop than with the person sitting in front of it. Finding out what's happening around the world is interesting, as is finding out what's happening around you. Right now the technology forces an unacceptable decision between doing one or the other.

As wearable computing wraps the Net around people this is going to change. I'm confident that for quite some time people will remain more engaging than computers, but finding the right person at the right time might require looking next door or on the next continent. Computers currently facilitate interactions that would not otherwise have occurred, with tools such as groupware that permit multiple people to work on a document, or videoconferencing that lets meetings happen among people at many sites. These functions require special software, hardware, and even rooms, and are consequently used primarily for narrow business purposes. Once wearables make it possible to see, or hear, anyone at anytime, then interacting with people physically far away can become as routine as talking to someone in front of you. A person having a polite conversation might equally well be talking to their neighbor, to a friend around the world, or even to their shoes; the whole notion of neighborhood becomes a logical rather than a physical concept.

"Networking" then becomes something much more compelling than passing around business cards at a cocktail party. If a group of people share their resources, they can accomplish many things that they could not do individually. For example, the performance of a portable radio is limited by the range of frequencies that are available (decided by the telecommunications regulatory agencies), the amount of noise in the receiver (which is ultimately determined by the temperature of the system), and the strength of the signal (a function of the size of the antenna). There isn't much that one person can do about these things, short of wearing cryogenic underwear or an inverted umbrella on his or her head. But radio astronomers have long understood how to combine the signals from multiple antennas to make an observation that could not be done by a single one. Processing the signals from many small antennas to effectively create one large one requires knowing precisely where each antenna is. This might not appear to work with people, who have an annoying habit of moving around, but we're learning how to design radios that can automatically synchronize and share their signals with other radios in their vicinity. If you want to exchange data with a wireless network from a wearable computer, and a few nearby people agree to let you temporarily run your signals through their radios, you can send and receive data twice as fast as you could alone. Since network traffic usually happens in bursts, when you're not using your radio they in turn can take advantage of it. The same thing can be done with computing, temporarily borrowing unused processor time from the people around you to speed up computationally intensive tasks.

Al Gore has observed that the U.S. Constitution can be viewed as a sophisticated program written for an enormous parallel computer, the populace. By following its instructions, many people together can do what they could not do separately. When computation becomes continuously available, the analogy becomes literal. Polling, voting, and legislating can become part of our ongoing collective communications rather than isolated events. Instead of using a few monitors to oversee a controversial election, an entire nation could bear witness to itself if enough people were wired with wearables.

It's not too far from there to see wearable computers as a new step in our evolution as a species. The organization of life has been defined by communications. It was a big advance for molecules to work together to form cells, for cells to work together to form animals, for animals to work together to form families, and for families to work together to form communities. Each of these steps, clearly part of the evolution of life, conferred important benefits that were of value to the species. Moving computing into clothing opens a new era in how we interact with each other, the defining characteristic of what it means to be human.

The Personal Fabricator

Thomas Watson, the chairman of IBM, observed in 1943 that "I think there is a world market for maybe five computers." In 1997 there were 80 million personal computers sold. To understand his impressive lack of vision, remember that early computers were

- large machines
- housed in specialized rooms
- used by skilled operators
- for fixed industrial operations
- with a limited market

From there it was too hard to conceive of a computer that could fit on a desk without crushing it, much less on a lap. Unseen beyond that horizon in packaging lay the revolutionary implications of personalization.

Once computers became small enough and cheap enough for individuals to own them, their application became a matter of personal preference rather than corporate policy. Big companies have an unerring knack for doing dumb things because so many people are involved in specifying and evaluating what someone else does that it's all too easy to forget to think. Once it became possible for individuals to write programs or configure packages to reflect their individual needs, then instead of marveling at someone else's stupidity they could do something about it. This represented a loss of control for the computer companies that were accustomed to prescribing what hardware and software their customers would use; in return, the market for computers grew to something more than five machines.

Now consider machine tools. These are the large mills and lathes and drills that are used in factories for fabrication. For the ordinary person, they are about as exciting as mainframe computers. In fact, they really are quite similar. Machine tools are

- large machines
- housed in specialized rooms
- used by skilled operators
- for fixed industrial operations
- with a limited market

Sound familiar? Big companies use big machines to make things we may not really want. Personal computing has not gone far enough: it lets us shape our digital environment, but not our physical environment. By giving computers the means to manipulate atoms as easily as they manipulate bits, we can bring the same kind of personalization to the rest of our lives. With the benefit of hindsight, plus a peek into the laboratory to see what's coming, this time around we can do better than Thomas Watson and recognize the impending arrival of the Personal Fabricator.

One of the eeriest experiences in my life came when I first opened the door of a 3D printer and took out a part that I had just seen on the screen of a computer. It violated the neat boundary between what is inside the computer and what is outside. In a strange way, holding the part felt almost like touching the soul of the machine.

A 3D printer is a computer peripheral like any other, but instead of putting ink on paper, or data on a disk, it puts materials together to make objects. Working with a 3D printer engages our visceral connection to the physical world, which has been off-limits for so long whenever a computer is involved. When I set up a 3D printer in my lab, people would come by just to touch the machine and the parts that it produced. Watching them, I could almost hear the mental gears grinding as they began to think about how they could make what they wanted, instead of purchasing what someone else thought they wanted. If a static shape can have that kind of impact, then I'm not sure how people will react when the output from the printer is able to walk out of it. Because we're also learning how to print sensors, motors, and logic.

To appreciate just how inevitable and important personal fabrication is, I had to retrace the history of computing in this new domain. When I arrived at MIT in the early '90s, you couldn't tell that it had been a pioneer in manufacturing technology. The few remaining machine shops on campus were in a sorry state, holdovers from the Iron Age in the midst of the Information Age. The campus had long since gone digital. Forlorn machine tools were left neglected by the throngs of students clustering around the latest computers.

As I set out to create a lab that could free computing from its confining boxes, I knew that we would need to be equally adept at shaping things as well as ideas. This meant that we would need to start with a good machine shop. A traditional machine shop is a wonderful place in which a skilled machinist can make almost anything. Since this one would be in the Media Lab, where neither the people nor the computers are traditional, this shop would need to be much more easily accessible.

The most versatile tool to be added was a milling machine. These come in two flavors. Manual ones are designed to be used by hand; a part to be machined is fixed to a bed that is moved under the rotating cutting tool by long lead screws turned by hand. Numerically controlled (NC) mills are designed to be run by a computer; the bed and sometimes the head are moved by motors under software control. The operator of an NC mill stands at a control panel that can be some distance from the workpiece, usually with a cryptic interface that makes it difficult to do much more than start and stop the machine.

I wanted both less and more than a conventional NC mill. Whatever we bought had to have a direct mechanical linkage that could be used manually. The feeling of the torque and vibration through the handles provides essential insight into how well a tool is cutting, which is invaluable for beginners and important also for experts working with unfamiliar materials. This is lost when the mill is controlled from a remote console.

Next, the mill needed to have a graphical interface that made it simple to specify shapes right on the machine, because the only things more unfriendly than typical NC controllers are the programs that engineers use to design parts. Finally, the mill had to be on speaking terms with the building's network, to make it easy to transfer designs from the many types of computers in use.

Machine tool distributors laughed when I described what I was looking for. I knew I had a problem when I got the same reaction from machine tool manufacturers. Over and over, I was told that I had to choose between the immediacy of a manual interface and the power of a numerical one.

As I traveled around the world, I started to make detours to visit manufacturers. The turning point in my quest came on a trip to see one of the largest European makers of machine tools. The day started with them proudly showing me their products—huge mills that cost $100,000, required skilled technicians to operate and maintain them, and were difficult to use for anything but repetitive operations. I then sat down with their eminent director of development to discuss my requirements. As I spoke, he became increasingly agitated. Soon he was pounding on the table and shouting that I must not ask for such a machine.

At first I wasn't sure whether to laugh or cry. As he continued to hold forth, I felt like I was being transported back a few decades to reenact a familiar scene. I wanted to buy a PC; he wanted to sell me a mainframe. He was making all of the same arguments for why small computers could never replace big computers: machining/computing requires expert skills, NC mills/mainframes have narrow markets, personal systems are toys that can't do serious work. We parted agreeing that we lived on different planets.

Fortunately, I came back and found what I was looking for in my backyard. A company on Boston's Route 128, Compumachine, started with a nice manual milling machine, added motors so that a computer as well as a person could control it, then put on a keyboard and screen to run it. Instead of the usual impenetrable industrial controller, the computer was a familiar PC.

In the machine tool industry this was viewed as dangerous lunacy, because the inevitable system crashes could cause real crashes of the mill, destroying it if not its operator. What makes the mill safe is an illusion: the computer only appears to be in charge. There's a layer of circuitry between it and the mill that actually issues the commands, monitoring everything the computer requests in order to prevent anything unsafe from happening.

An NC mill is such a specialized piece of equipment that it usually has to earn its keep with manufacturing operations. Putting one in the hands of graphic designers, and programmers, and musicians, led to all sorts of clever things getting made in a way that would never occur to a traditional engineer. The mill was used to build a parallel computer with processors embedded in triangular tiles that could be snapped together in 2D or 3D sculptures, exchanging data and power through the connections. Miniature architectural models were created to serve as tangible icons for a computer interface. The success of these kinds of projects led me to wonder if physical fabrication could be made still simpler to reach still more people.

An alternative to making something by machining away parts that you don't want is to assemble it by adding parts that you do. Lego blocks are a familiar example of additive fabrication that make it possible to quickly and easily build impressive structures. There's a surprising amount of wisdom in this apparently simple system; Lego has spent decades getting the bricks just right, experimenting with the size of the posts, the angle of the faces, the hardness and finish of the plastic. The earliest bricks look almost like the current bricks, but these ongoing incremental improvements are what make them so satisfying to play with.

Anyone who has been around kids building with Legos understands the power of a child's drive to create. This is on display at the Legoland near Lego's headquarters in Denmark, an amusement park filled with spectacular Lego creations. There are Lego cities with buildings bigger than the kids admiring them, giant Lego animals, Lego mountain ranges. What's so amazing about this place is the kids' reaction. There are none of the things we've come to expect that children need to be entertained, no whizzy rides or flashy graphics or omnipresent soundtracks. Just great structures. And the kids are more engaged than any group I've ever seen, spending awed hours professionally appraising the marvels on display. The people who work for Lego, and who play with Lego, share a deep aesthetic sense of the pleasure of a nicely crafted structure.

In the Media Lab, Professor Mitch Resnick has one of the world's most extensive Lego collections. For years his group has been working to extend the domain of Lego from form to function, embedding sensors and actuators as well as devices for computing and communications, so that the bricks can act and react. At first, conventional computers were externally interfaced to the Lego set; now they've been shrunken down to fit into a brick. These let children become mechanical engineers, and programmers, with the same comfortable interface they've known for years. Kids have used the system to make creatures that can dance and interact, to animate their fantasy worlds, and to automate their homes. This system has been so successful that it has now left the lab and become a new line of products for Lego. The only surprise, which isn't really a surprise, is that grown-ups are buying the sets for themselves as well as for their kids.

At the opposite end of the spectrum is Festo. Festo is to industrial engineers as Lego is to kids; they make the actuators and controllers that are used to build factories. In fact, Lego's factories are full of Festo parts making the Lego parts. Festo also has a system for prototyping new factories and teaching people to design and run them. This is Lego for grown-ups. When I first brought its components into my lab there was a rush of ooh's and aah's, because it could do what Lego couldn't: make large precise structures. The hefty, shiny metal parts spoke to their serious purpose.

Following detailed instructions, with some effort, we used this system to put together an assembly line to make pencil sharpeners. And that's all we did. People left as quickly as they appeared, because it soon became clear that it was too hard to play with something this exact. Too many specialized parts are needed to do any one thing, assembling them is too much work, and interactively programming the industrial controllers is all but hopeless.

Between Lego and Festo the needs of children and industrial engineers are covered. Mitch and I began to wonder about everyone else. Most people don't make most of the things that they use, instead choosing products from a menu selected for them by other people. Might it be possible to create a system that would let ordinary people build things they cared about? We called this "Any Thing" and set up a project to develop it. Drawing on the lessons from Lego and Festo, this kit would build electrical connections for data and power into every mechanical joint, so that these capabilities did not need to be added on later. The parts would snap together in precise configurations, without requiring tools. And it would be made out of new composite materials, to be light, strong, and cheap. Armed with these guiding principles, we first tried out computer models of the parts, then prepared to start building it. At this point we called a design review meeting. We filled a room with as many alternative construction kits as we could find for comparison, and then added as many people and as much pizza as we could fit in. The night took many surprising turns.

Unfortunately, Any Thing had been developed entirely by a bunch of geeky guys (like me). This was the first time any women had seen it. One of the first questions that one of my female colleagues asked was what it was going to feel like; our jaws dropped when we realized that we had designed bones without thinking about skin. As the night proceeded and people played with the various commercial construction systems, they were uniformly frustrated by the nuisance of chasing around to find all of the fiddly little bits they needed to complete whatever they were working on. The Any Thing vision of personal fabrication necessarily entailed a personal warehouse to store an inventory of all the required components.

The more we talked, the more we realized that the most interesting part of Any Thing was not our proposed design, it was the 3D printer we were using to prototype it. Three-dimensional printing takes additive fabrication to its logical conclusion. Machine tools make parts by milling, drilling, cutting, turning, grinding, sanding away unwanted material, requiring many separate operations and a lot of waste. Just as a Lego set is used to construct something by adding bricks, a 3D printer also builds parts by adding material where it is needed instead of subtracting it where it is not. But where Lego molds the raw materials to mass-produce bricks in the factory, in a 3D printer the raw materials are formed on demand right in the machine.

There are many ways to do this. One technology aims a laser into a vat of epoxy to cure it just where it is needed, another spreads layers of powder and squirts a binder where it should stick together, and our Stratasys 3D printer extrudes out a bead of material like a computer-controlled glue gun that leaves a three-dimensional trail. These techniques can all go from a computer model to something that you can hold, without requiring any special operator training or setup. This convenience is significantly decreasing the time it takes companies to make product prototypes. The machines are currently expensive and slow, but like any promising machinist they are steadily becoming more efficient.

As interesting as 3D printing is, it's still like using a PC to execute a mainframe program. The end result is not very different from what can be made with conventional machine tools, it's just the path to get there that is simpler. The real question posed by our Any Thing design review was whether 3D printing could be extended as Lego had been to incorporate sensors and actuators, computing and communications. If we could do this, then instead of forcing people to use their infinitely flexible and personal computers to browse through catalogs reflecting someone else's guesses at what they want, they could directly output it. Appliances with left-handed controls, or ones with easy-to-read large type, would become matters of personal expression rather than idle wishes.

This is the dream of the personal fabricator, the PF, the missing mate to the PC. It would be the one machine that could make all your others, a practical embodiment of the perennial science-fiction staple of a universal matter output device. As we began to dare to dream of developing such a thing, we realized that we had been working on it all along.

Joe Jacobson's printer for the electronic book project was already laying down the conductors, insulators, and semiconductors needed to make circuits on paper. Just as an ink-jet printer has cartridges with different-colored inks, it is possible to provide a printer with more types of input materials so that it can also deposit structural shapes and active elements. One of the first successes was a printed motor that used carefully timed electric fields to move a piece of paper. If your desk was covered by such a structure then it could file your work away for you at the end of the day.

The semiconductor industry is trying to reach this goal of integrating sensors, circuits, and actuators, by cutting out tiny little silicon machines with the fabrication processes that are currently used to make integrated circuits. This idea is called MEMS, Micro-Electro-Mechanical Systems. Although the industry sees this as an area with great growth prospects, MEMS fabrication has the same problem as mainframes and machine tools. MEMS devices are designed by specialists and then mass-produced using expensive equipment. The desktop version could be called PEMS, Printed Electro-Mechanical Systems. Unlike MEMS, it can make people-size objects rather than microscopic machines, and the things are made where and when they are needed rather than being distributed from a factory.

A PEMS printer requires two kinds of inputs, atoms and bits. Along with the raw materials must come the instructions for how they are to be placed. One possibility is for these to be distributed over the Net. Instead of downloading an applet to run in a Web browser, a "fabblet" could be downloaded to the printer to specify an object. This would significantly lower the threshold for a company or an individual to sell a new product, since all that gets sent is the information about how to make it. But it stills leaves people selecting rather than designing what they want.

To bring engineering design into the home, the CAD software that engineers currently use is a valuable guide for what not to do. These packages generally betray their mainframe legacy with opaque user interfaces and awkward installation procedures, and isolate the design of mechanical structures, electrical circuits, and computer programs into modules that are barely on speaking terms.

A design environment to be used in the home should be so simple that even a child could understand it. Because then there's some hope that their parents might be able to also. There's already a pretty good precedent that's routinely used by children without any formal training: Lego. It provides simple standard parts that can be used to assemble complex systems.

That was the original inspiration for the Any Thing project, which in retrospect was the right idea for the wrong medium. It was a mistake to assume that the physical fabrication process had to mirror the logical design process. By moving the system from a bin of parts to a software version that exists only inside a computer program, the intuitive attraction of building with reusable components can be retained without needing to worry about ever running out of anything. Computer programmers use "object-oriented programming" to write software this way; with a PEMS printer the objects can become tangible.

It's important that the reusability be extended to the physical as well as the virtual objects. The arrival of fast high-resolution printers has created the paperfull office of the future because printing is now so convenient. This is a cautionary tale for PEMS, which threatens to do the same for physical things. Two-dimensional sheets of paper get thrown away in three-dimensional recycling

bins; three-dimensional objects would need a four-dimensional recycling bin.

A slightly more practical alternative is to let the PEMS printer
separate discarded objects back into raw materials. This is what a
state-of-the-art waste facility does on a large scale; it's a bit easier
on the desktop because the task is restricted to a much smaller
group of materials that can be selected in advance to simplify their
separation. Just as coins get sorted by weight, differences in the materials' density, melting temperature, electrical conductivity, and so
forth can be used to separate them back into the printer's input
bins.

Complementing a PEMS printer with intuitive design tools and
the means to recycle parts provides all of the ingredients needed to
personalize fabrication. This just might eliminate some of the many
bad ideas that should never have been turned into products. I have
a theory for why so many companies full of smart people persist in
doing so many dumb things. Each person has some external bandwidth for communicating with other people, and some internal
processing power for thinking. Since these are finite resources, doing more of one ultimately has to come at the expense of the other.
As an organization expands, the volume of people inside the company grows faster than the surface area exposed to the outside
world. This means that more and more of people's time gets tied up
in internal message passing, eventually crossing a threshold beyond
which no one is able to think, or look around, because they have to
answer their e-mail, or write a progress report, or attend a meeting,
or review a proposal. Just like a black hole that traps light inside,
the company traps ideas inside organizational boundaries. Stephen
Hawking showed that some light can sneak out of a black hole by
being created right at the boundary with the rest of the world;
common sense is left to do something similar in big companies.

So many people are needed in a company because making a

product increasingly requires a strategy group to decide what to do, electrical engineers to design circuits that get programmed by computer scientists, mechanical engineers to package the thing, industrial engineers to figure out how to produce it, marketers to sell it, and finally a legal team to protect everyone else from what they've just done. It's exceedingly difficult for one person's vision to carry through all that. Focus groups help companies figure out when they've done something dumb, but they can't substitute for a personal vision since people can't ask for things they can't conceive of.

Companies are trying to flatten their organizational hierarchies and move more decision making out from the center by deploying personal computing to help connect and enable employees. But the impact of information technology will always be bounded if the means of production are still locked away like the mainframes used to be. For companies looking to foster innovation, for people looking to create rather than just consume the things around them, it's not enough to stop with a Web browser and an on-line catalog. Fabrication as well as computing must come to the desktop.

The parallels between the promise and the problems of mainframes and machine tools are too great to ignore. The personal fabricator now looks to be as inevitable as the personal computer was a few decades ago. Adding an extra dimension to a computer's output, from 2D to 3D, will open up a new dimension in how we live.

This revolutionary idea is really just the ultimate expression of the initial impetus behind multimedia. Multimedia has come to mean adding audio and video to a computer's output, but that's an awfully narrow notion of what constitutes "media." The images and motion don't have to be confined to a computer screen; they can come out here where we are.

Smart Money

Barings Bank was founded in 1762. In its long history it helped to finance the Louisiana Purchase (providing money Napoleon needed to keep fighting his wars), and counted the Queen among its loyal customers. In January of 1995 a twenty-eight-year-old trader for Barings in Singapore, Nick Leeson, lost most of what eventually proved to be $1.4 billion by trading futures in the Japanese Nikkei Index. That was twice the bank's available capital; by February the bank had folded, and in March it was sold to the Dutch bank ING for £1.

In July of that same year, Toshihide Iguchi, a trader for Daiwa Bank in New York, confessed to the bank that he had lost $1.1 billion trading apparently harmless U.S. Treasury securities. By November Daiwa was forced out of the United States for concealing the losses.

The next year Yasuo Hamanaka, a copper trader for Sumitomo, dwarfed these scandals by admitting to having lost $2.6 billion in copper trading. In the preceding years he had single-handedly kept the copper market artificially inflated, leading to the digging of unnecessary new mines and a subsequent glut of copper. After his manipulations were revealed the price collapsed by 25 percent.

Losing large sums of money no longer requires even the specialized skills of a rogue trader; Robert Citron, the duly elected treasurer of Orange County, California, lost $1.7 billion by investing in derivatives. This led to the largest municipal default in U.S. history, by one of its richest counties, no less than the birthplace of Richard Nixon.

As quickly as enormous amounts of money are vanishing into economic thin air they are reappearing elsewhere. Markets are valuing bits far more than atoms. In 1997 U.S. Steel had twenty thousand employees, sales of $6.8 billion, and assets of $6.7 billion. The total value of U.S. Steel stock was $2.9 billion. Last year Yahoo, the company with the popular Internet index site, had just 155 employees, sales of $67 million, and assets of $125 million. But the total value of Yahoo stock was about the same as U.S. Steel, $2.8 billion. Netscape already had kicked off the frenzy of public offerings of stock in Internet companies, with a first day of trading that brought the net worth of the company to $2.2 billion.

The market's conversion rate between atoms and bits has been running at about ten to one. The stock market values Ford and GM—which together have a million employees and assets of $500 billion—at $100 billion. The stock for Microsoft and Intel, which jointly have about a hundred thousand employees and assets of $50 billion, is worth $300 billion.

The distribution of this new wealth is distinctly uneven. According to the Census Bureau, in 1968 in the United States the top 5 percent of households received 16.6 percent of the total household income; by 1994 that had climbed to 21.2 percent. Similar redistributions are happening globally: World Bank numbers show that poor- and medium-income countries held 23 percent of the world's wealth in 1980, dropping to 18 percent eight years later.

There's something happening here. The economy appears to be

divorced from the laws of physics, creating and eliminating great wealth with little apparent connection to the material world. And that is exactly the case: money is increasingly digital, packets of data circulating in global networks. The electronic economy is weaving the digital and physical worlds together faster than anything else. The demand for more sophisticated ways to manipulate money is forcing computing to go where money goes, whether in a trader's workstation, a smart card in a wallet, or a thinking vending machine.

Freeing money from its legacy as a tangible asset carries with it great promise to make the economy more accessible to more people, but we'll be in trouble if we continue to act as if money is worth something. The bits of electronic cash still retain a vestigial reflection of their origin in the atoms of scarce resources.

Just after World War II, the Bretton Woods Agreement fixed a conversion rate between dollars and gold at $35 an ounce, and in turn set exchange rates between the dollar and other currencies. The U.S. government was obligated to convert dollars into gold upon request, using the cache held at Fort Knox. In 1971, saddled with a dwindling gold supply, a persistent recession, and an expensive war in Vietnam, President Nixon took the dollar off the gold standard. Since then, the value of currencies has been fixed solely by what the markets think they are worth. We maintain a polite fiction that money stands for a valuable commodity, but it now represents nothing more than beliefs.

The implications of the end of the gold standard reach far beyond the passionate debates of that time, because of an apparently unrelated event that was quietly happening in a handful of laboratories: in 1971 the ARPANET entered into regular service. This was a project supported by the U.S. Defense Department Advanced Research Projects Agency (DARPA) to create a computer network that had no central control and multiple ways to route messages, just the thing to survive a nuclear attack. In 1971 it had all of fifteen host machines. Its original research function was soon forgotten as the network filled with people sending e-mail, and connecting more computers. With millions of hosts on it by 1989, DARPA gave up on trying to manage the experiment, and the Internet was born.

At the end of the gold standard, computing was done by relatively isolated mainframes; computers were needed to record financial trends but did not create them. With the arrival of instantaneous global connectivity and distributed computing, money can now travel at the speed of light, anywhere, anytime, changing its very nature. The only physical difference between a million and a billion electronic dollars is the storage of ten extra bits to fit the zeros; money has become one more kind of digital information.

The apparent function of money is clear. It has a transferable value; I can exchange a dollar bill for a carton of milk because the grocer can then exchange the dollar for something else of comparable value. An electronic representation alone does not change this. Digital money has been around for years in the form of electronic funds transfers done by financial institutions, transactions that differ in scale but not in kind from buying milk. They could in principle be done equally well by moving around (very) large amounts of physical currency. The spread of the Internet has now brought electronic commerce into the home. The grocer can send a packet of data describing the milk to your Web browser, and you can send a packet of data back to the grocer with your credit card information.

There's a popular misconception that it is dangerous to transmit a credit card number over the Internet. This dates back to the time when network traffic was not encrypted, so that a malicious or even just curious person with the right software could connect to a network and read the contents of the packets sailing by. This was an eavesdropper's dream, because the data came to you instead of your having to go to it. Although I've had taxi drivers lecture me on the dangers of the Internet, the reality now is that there are solid cryptographic protocols that make this kind of network sniffing impossible. In a "public-key" cryptosystem you can openly advertise an ID much like a phone number that lets people easily encode a message before sending it to you, but that makes it impossible for anyone other than you to decode it unless they know your secret key associated with the public key. This can be done by using what are called secure socket network protocols, built into most Web browsers.

Modern cryptography goes much further, splitting hairs that you might not have realized you had. If you don't trust the grocer, a "zero-knowledge proof" lets you give the grocer enough information to be able to bill you without letting the grocer learn either who you are (so you don't have to worry about getting junk mail) or what your account is (so you don't have to worry about later unauthorized charges). Conversely, by using a digital signature, the grocer can prove that an advertisement came from the store and not someone trying to spoof store business. Electronic certificates permit a third party whom you trust to remotely vouch for the grocer's reliability before you sign a contract for milk delivery.

Smart cards build these cryptographic capabilities into your wallet instead of your Web browser. A smart card looks like a credit card but acts like a dollar bill. A chip in the card stores electronic money that can be spent directly, unlike the account number stored in a credit card that needs access to a central computer to be used. The first widespread use of smart cards came in pay phones in Europe, where they eliminated the need to fumble through a Pocket-of-Babel's worth of change to make a local phone call. They're now expanding into retailing so that both buyers and sellers can stop carting around currency.

After all, what could be more useless than a penny? They travel between bowls on shop counters and jars on bureaus because it's too much trouble to take them back to the bank. There's a solid economic argument that the net cost of handling pennies exceeds their worth, therefore they should be eliminated as the smallest currency unit. The future of the penny does not look bright.

Qualify that: the future of a *physical* penny does not look bright. The value of *electronic* pennies is just beginning to be appreciated. On the Internet the packets of data representing financial exchanges are a drop in the bucket compared to high-bandwidth media such as video. There's plenty of room to send much more of them. The total amount of money in circulation can't go up, but the size of a transaction can certainly go down. The overhead in selling a product sets a floor on what people are willing to purchase. Either you buy a CD, or you don't. If, however, the CD arrives as bits on the network, you could pay for it a track at a time. Rather than buying an expensive video game cartridge, you could be billed for each level of the game as you play it. Your taxes that pay for garbage collection, or highway maintenance, could be collected each time you take out the trash or drive your car. The incentive for a consumer to consider buying this way is to be able pay for just what gets used rather than paying in advance for the expected lifetime cost of something. The advantage for a merchant is the opening up of new markets by decreasing initial purchase costs, the ongoing revenue stream, and the detailed and continuous usage information that gets provided (if the buyer agrees to make that available).

As interesting as it is to introduce micropayments and cryptography into commerce, these are just ways to repackage our old model of money, analogous to your buying milk in lots of tiny containers, or visiting the store with a paper bag over your head to protect your privacy. Most electronic representations of old media start out by emulating their functionality and then later generalize beyond the hidden assumptions. For example, the first formats for digital pictures stored an array of intensities to match each part of an image, before people realized that a picture could be stored in order of importance so that transmitting a small amount of data provides enough information to produce a small image that can be refined as more information is received, or identifying data can be hidden in the image to catch forgery, or the parts of a picture can be encoded separately so that they can later be rearranged. The familiar notion of a picture has grown to encompass many new functions. Likewise, moving money as bits instead of atoms creates an opportunity to expand its identity far beyond the familiar forms that it has taken in the long history of commerce.

Each day, over a trillion dollars circulates in the global currency market. That's a lot of money. The bulk of this activity has nothing to do with getting francs for a trip to Paris; it is a massive hunt for profit and security in the inefficiencies of the market. Now this is very different from buying a gallon of milk. There is no intention to use what is bought or sold; the real content of a currency trade is the information that motivates it. What's being exchanged are opinions about whether the yen is overvalued, say, or whether the Fed will lower interest rates. Financial speculation isn't a new idea, but the size is. The sums need to be so vast because the trades are exploiting minuscule differences in the value of currencies that show up only on the scale of a national economy. And they can be so big because the combination of an electronic representation of money with a global computer network makes taking a $100 million foreign exchange position no more difficult than sending an e-mail message.

The dollars used in a large financial trade have a very different character from the dollars used to buy a gallon of milk, but we continue to presume that they are identical. The dollars that you use to purchase milk are useful in small quantities, and essentially all of their value goes into paying for the milk (there's just a bit of extra demographic information that the seller can collect about where and when you bought the milk). The dollars that a trader uses are useful only in big quantities because of the small margins and the large minimum size of the trades, and there's no desire to actually use what is being bought. This distinction can be thought of as atom-dollars and bit-dollars.

People, and countries, that don't have access to bit-dollars are increasingly being left behind, unable to compete in the digitally accelerated economy. The corner grocer has to move a lifetime of milk to match the resources committed by a few keystrokes on any trader's keyboard. And people who can participate in the new economy are seeing bigger and faster swings between spectacular riches and ruin because normal economic fluctuations can be magnified almost impossibly large.

Just as it was once thought that printed dollars needed to be backed up by a gold standard, we now assume that bit-dollars must be interchangeable with atom-dollars. The end of the gold standard was an economic leap of faith that remains controversial, but it arguably is what enabled the rapid ensuing economic development in many parts of the world. Now, given the spread of computation and a digital representation for money, the conversion between electronic and tangible assets can depend on the nature of a transaction. It's possible to end the atom standard.

We already accept that the grocer cannot receive a dollar for a carton of milk and then directly exchange it for a currency position, because of the large minimums and rules for foreign exchange market access. The converse would be to have $1 million made in a currency transaction buy something less than a million gallons of milk. This could be implemented by making the value of a dollar depend on its context. Its purchasing power could vary based on how many dollars it was earned with and then how many dollars it is being spent with.

The $1 million earned in a trade could still buy a million-dollar dairy, but for individual purchases at the corner market it would be worth less than dollars that had been earned more slowly in smaller quantities. Such a scaling is not a simple devaluation or transfer of wealth from the rich to the poor; it's a recognition that the money is being used in entirely different ways and it is not at all obvious that the conversion factor should be one-to-one. Considering such a change raises enormous economic, political, and social questions, but now that it is technically possible to do it, choosing not to is itself an even more significant decision.

There's a precedent for valuing something based on a rule that depends on its history: that's what derivatives do, the new financial instruments that are now so painfully familiar to Orange County. A derivative sets a price based on the behavior over time of an asset, rather than the worth of the underlying asset itself. For example, an option could be designed that would pay off if the activity in a given market increases, regardless of which way the market moves. These instruments were developed because the markets are so random that it is all but nonsensical to forecast the daily values but quite reasonable to forecast changes in attributes such as the volume of trades. What's interesting about options is this introduction of algorithms to the price of something, so that value can be associated with change.

Derivatives themselves are not dangerous. The reason that they are associated with so many billion dollar boo-boos is that it's not possible to assess their present worth, or your vulnerability to changes in the market, simply by tallying up an account balance. In a financial firm the traders are at the top of the pecking order, surrounded by the most advanced computers and data analysis techniques to guide and make trades. Then there's a risk assessment group that trails after them trying to determine the firm's exposure to risk. The latter is generally a sleepy enterprise located far from the trading floor, lacking the tools of the former. This imbalance is a recipe for disaster, because committing resources to a position is easy compared to evaluating the financial risk of a portfolio of derivatives, which requires a detailed understanding of the state of the markets and how they are likely to move. In this unequal competition, the trading side almost always wins out over the risk assessment side. Some firms effectively give up and ask the traders to evaluate themselves. Tellingly, Leeson at Barings, and Iguchi at Daiwa, both kept their own books because their managers from an earlier generation did not feel capable of supervising them.

The solution is to recognize that the further that assets get divorced from underlying resources, the more necessary it becomes to merge spending with monitoring. Each new algorithm for valuing something must be mated to a new algorithm for assessing it. Back-end accounting then moves to the front lines, drawing on the same data feeds and mathematical models that are used for trading. Too many people still think of smart money in dumb terms, assuming that like gold bars it has an intrinsic worth that won't change very much if you look away from it. Understanding the implications of a derivative outstrips human intuition, particularly as computers, instead of people, increasingly initiate trades. Since physicists have spent centuries trying with limited success to understand the behavior of complex interacting random systems, it's too much to expect one trader to be able to do so (the physicists on Wall Street certainly haven't been able to). The money must do a better job of supervising itself. This is a cultural problem more than a technological one, requiring an acceptance that the bits representing money should come packaged with bits representing algorithms for tracking their worth.

Stepping back to recognize that it is more natural to view electronic money as comprising the combination of data describing quantity and algorithms specifying valuation, it then becomes possible to create combinations that are useful for purposes far from high finance. Think what could happen if e-cash contained the means for ordinary people to add rules. A child's allowance might be paid in dollars that gain value based on the stability of the child's account balance, to make the benefits of saving apparent in the short term instead of just through the long-term accrual of interest. Or a store's price guarantee could be implemented by pricing an item in dollars that stay active after the transaction, and that have a value derived from the market for that item. You don't need to search for the lowest price; the money can. If a competing car dealer lowers the price on a car the day after you bought one at another dealer across town, that information can automatically initiate a transfer between your dealer's bank account and yours to refund the difference. The dealer gets your business if it thinks it can always beat the competition; you get the freedom from worrying about finding the best deal.

If money can contain algorithms then this kind of oversight could be built into every transaction rather than being a matter of fiscal policy alone. Central banks have just a few coarse levers with which to manage a nation's economy, primarily controlling the prime interest rate that the government charges banks. If it's low, new money gets printed at less than its real worth, in effect making a loan from the future to the present. That may be a good idea in a depression, but otherwise it saddles your children with your debt. If the rate is high, then money costs too much and people are forced to sacrifice now for the sake of saving for the future. This one number applies to everyone, everywhere.

Once money is manipulated by machines, then it can be personalized to reflect both national and local interests. Dollars could have a latency period when they're not valid after transactions, to slow down panic spending. In a time of economic turmoil this would mean that you might have to wait a day after withdrawing your life savings from a bank before you could put all the money into a get-rich-quick scheme. To encourage regional economies, dollars could have a value that decreases with distance from their origin. Boston dollars would not go as far in Texas, and vice versa. Or, to reflect true environmental costs, dollars might be worth less when purchasing nonrenewable resources than renewable ones. These are not matters to be settled once for everyone everywhere; they are dynamic decisions to be made by buyers and sellers on an ongoing basis.

Such goals are currently implemented by familiar mechanisms such as tax policies and sales contracts, without needing something so ambitious as a redefinition of the nature of money. But as the money, and the information about how the money can be and is being spent, merge in packets of economic information running around networks, the distinction between the money and the supporting information becomes less and less meaningful, until it becomes simpler to recognize the financial data packet as a new definition of money.

Just like any other digital media, for this to succeed there must be a vigorous community of money developers working with open standards. If any economic or technological lesson has been learned over the last few hundred years, it's that central planning cannot substitute for the creative anarchy of relatively free markets. It would be a shame if the dollar bill had to relearn all of the mistakes of previous computer standards, which have a nasty habit of rediscovering old errors and not anticipating future growth. A number of computer operating systems started out using 16-bit addresses to identify memory locations. This is enough to use 65,536 addresses, which at one time seemed unlimited. As hardware and software both expanded beyond this boundary, awkward patches were needed to use some of the precious addresses to point to other blocks of addresses that could be swapped in and out as needed. Because the number of address bits was fixed by the operating system, this problem continued to recur long after it was clear that sixteen bits wasn't adequate. Similarly, most every network standard (including the IP protocol that runs the Internet) has run out of addresses and needs a redesign. The solution to this problem of lack of foresight is not better foresight, it is more humility. Since it's not possible to anticipate future growth, standards are increasingly designed to let independent developers add new capabilities. One reason that the computer game Doom has been so popular is that the authors released it with a specification that lets other people add new worlds and new capabilities to it. There's a lively little industry that has grown up around the game, continually reinventing it in a way that one team of programmers alone could never do.

Opening up fiscal policy to the people who brought us Doom is admittedly a terrifying prospect, but it's instructive to compare the history of the Internet and Bitnet. In the dark ages of computing one didn't send e-mail, one "sent a Bitnet." The Bitnet was an international network of mainframes organized along the line of the Soviet economy with central planning and management; the Internet was (and still is) a relatively anarchic collection of distributed switches and collectively supported evolving protocols. Guess which one survived? Bitnet was shut down in 1996, a centralized monoculture that could not keep up with the frenetic growth and distributed innovation of the Internet. Even with the occasional glitches, the Internet is perhaps the largest and most reliable system ever constructed, precisely because there is no central location to fail or mismanage and so many people can contribute to its development.

The history of computing is littered with unsuccessful standards that sought reliability through specification rather than experience.

When the car companies and other heavy industries wanted to install computer networks in their factories they decided that what everyone else was using was not reliable enough, so they came up with their own standard that was supposed to be foolproof. They now use the open standards for networking that everyone else does. They found that reliability depends much more on having a diverse community working on developing and deploying the networks than on trying to control their design. As a result, intelligence is now distributed throughout factories rather than being centralized in a few computers, or people.

Conversely, there are many bad standards that have been very successful because they came along at the right time. If a standard gets specified too soon, it reflects early ignorance rather than useful experience. If it comes too late, it's irrelevant no matter how good it is. In between is a time when a standard appears too soon to really know how to write it well, but soon enough to be able to have an impact. MIDI, the protocol for connecting electronic musical instruments, is a good example of a bad standard (because it can't scale up to handle installations larger than what was originally envisioned), but it arrived at just the right moment to create a new industry. You can hear the result in almost any popular music recording.

The economy is now ready for something similar. Just as technological improvements in the tools to create sounds or send messages have led to a transformation of our notion of what it means to make music or interact with others, the increasing manipulation of money by machines rather than people presents an opportunity to transform the global economy to better reflect personal needs. It's not yet clear exactly how to write an open standard for smart money that merges value with behavior, but markets are already doing this imperfectly, and history provides strong guidance for how to go about creating standards for reliable, scalable, digital systems. These are increasingly questions for programmers, not economists.

Commerce is where the physical world and the digital world will always meet. A purchase must become embodied to be usable. As electronic information appears in more and more of those transactions, it is essential that money be freed to complete its journey from a tangible asset to a virtual concept. Buying and selling is increasingly mediated by smart systems that still use dumb money. Abstracting money from its legacy as a thing will make it easier for still smarter things to manipulate it, helping them better act on our desires.

WHY

. . . should things think?

the rights of people are routinely
infringed by things, and
vice versa

✦

dumb computers can't be fixed
by smart descriptions alone

✦

useful machine intelligence requires
experience as well as
reasoning

✦

we need to be able to use all of our
senses to make sense
of the world

Rights and Responsibilities

Telemarketing. Thirteen unlucky letters that can inflame the most ill-tempered rage in otherwise well-behaved people. The modern bane of dinnertime: "click . . . uhh . . . Hello . . . Mr. Gersdenfull, how are you today?" Much worse than I was a few minutes ago. The most private times and places are invaded by calls flogging goods that have an unblemished record of being irrelevant, useless, or suspicious. Answering these calls is a pointless interruption, listening to the answering machine pick them up is almost as bad, and shutting off the phone eliminates the calls that I do care about.

Who's responsible for this unhappy state of affairs? I don't blame the poor person at the other end of the line who is stuck trying to make a living by being subjected to other people's ire. And I don't really fault their bosses, who are doing what they are supposed to do in a free market, taking advantage of a perceived commercial opportunity. I blame the telephone, and Pope Leo X.

If modern information technology dates from Gutenberg's development of movable metal type in the 1450s, then Martin Luther was the first master of the new medium, and Leo X the first opponent of its personalization. At the time in Europe the Church had a monopoly on divine communication, which had to use its format of a Latin Bible, available only through the word of priests who were its authorized trained representatives. Popes Julius II and then Leo X took advantage of the monopoly to fund an ambitious building program for St. Peter's Cathedral by selling indulgences. These handy letters offered the remission of sins, and a shorter period in purgatory along the way to guaranteed eternal salvation. Johann Tetzel, the monk whose territory for selling indulgences included Luther's community, was particularly creative in marketing them. On his convenient price list murder cost eight ducats; witchcraft was a bargain at two ducats. He even offered indulgences to help save family and friends who had already passed away. Such a deal.

Luther's disgust at this venal corruption of the principles of the Church drove him in 1517 to post a complaint on the local BBS of the day, the door of the castle church in Wittenberg. His Ninety-five Theses (Luther 95?) argued that salvation must come from personal faith and deeds, not purchase of papal authority. His story might have ended there, but for a difference in responding to the affordances of a new medium. Luther recognized that the printing press enabled one-to-many communications. Before movable metal type, hand-copied books had been so expensive that only the wealthy or the Church had direct access to them; by 1500 there were millions of printed books in circulation. Within weeks of its original posting, Luther's Theses had been copied and spread throughout Europe. Leo in turn understood the threat that widespread printing posed to the authority of the Church. He tried to restore central control by issuing a papal bull to require certification from the Church before book publication, and in 1521 excommunicated Luther, forbidding the printing, selling, reading, or quoting of his writings. After Luther responded by burning the bull of excommunication, Leo had him summoned for judgment to the tastefully named Diet of Worms, where he was exiled.

Luther won. His brief exile gave him a chance to translate the

New Testament into German, and his excommunication gave his book sales a great boost, bending the Church's propaganda apparatus into creating PR for him. The Reformation sprang from Luther's Theses. He and his fellow authors understood that they were writing for the printing press and hence produced short tracts that were easily duplicated—the first sound bites. The authors of the Counter-Reformation didn't appreciate this change, writing hefty tomes in the old style that were hard to distribute.

The Reformation message fell on the fertile soil of a laity fed up with a clergy that forced divine interactions through their preferred means, demanded meaningless ritual observances, and introduced self-serving modifications into the liturgy. The availability of vernacular (non-Latin) Bibles let ordinary people evaluate and then question the claim of privileged papal authority. From here grew the modern concept of personal freedoms distinct from communal responsibility.

Fast-forward now a few centuries. I had just received a new laptop, one of the most powerful models of its generation. I wanted to use it to make a presentation, but for the life of me I couldn't find any way to turn on the external video connection. Resorting to the ignominy of reading the manual, I found this gem of technical writing explaining how to get into the machine's setup program to switch the video output:

```
1. Make sure that the computer is turned off.
2. Remove any diskette from the diskette drive.
3. Turn on the computer.
4. Watch closely the flashing cursor in the top-
   left corner of the screen. When the cursor
   jumps to the top-right corner of the screen,
   press and hold Ctrl+Alt, then press Insert.
   You must do this while the cursor is at the
   top-right of the screen. Release the keys.
```

Performing this feat requires three fingers and split-second timing. Even the Roman Catholic Church couldn't top this for imposing nonsensical rituals onto sane people. And, like the Church, the computer industry forces most people's interactions to go through one approved means (windows, keyboard, mouse), and it sells salvation through upgrades that are better at generating revenue than delivering observable benefits in this lifetime. The Reformation was a revolt, enabled by the spread of information technology, against the tyranny of central authority. I'm increasingly concerned about the tyranny of the information technology itself. In many ways we're still fighting the Reformation battle to establish access to printing presses, when we should be worrying about the access of the printing presses to us.

Luther's Ninety-five Theses and Leo's response helped to inspire the English Bill of Rights after King James II was deposed in 1688. Among other sins, James had abused the ultimate line-item veto, the royal Dispensing Power that let him make exceptions to laws, to try to place Catholics throughout the government and military. In 1689 his Protestant son-in-law and daughter, William of Orange and Mary, were crowned King and Queen (thanks, Dad). Parliament handled the sticky question of the authority to do this by presenting William and Mary with what became the Bill of Rights. This established the grounds upon which James was deemed to have forfeited his right to rule, and then took the remarkable step of listing thirteen new rights that the monarchy would have to accept, including freedom of speech in the Parliament.

The U.S. Constitution in 1787 did not explicitly include a Bill of Rights, a source of widespread dissatisfaction. People didn't trust the Federalist argument that they had nothing to worry about since the Constitution didn't give the government authority over individual rights. This was remedied by the first act of the first Congress, passing the amendments that were ratified by the states in 1791 to create the U.S. Bill of Rights, modeled on and in some places copied from the English Bill of Rights. The First Amendment, of course, extended the freedom of speech and the press to everyone.

The importance of widespread access to the means of communications was echoed in the Communications Act of 1934, the bill that established the FCC to regulate the emerging radio and telephone services. It states its purpose to be ". . . to make available, so far as possible, to all the people of the United States a rapid, efficient, Nation-wide, and world-wide wire and radio communication service with adequate facilities at reasonable charges. . . ." The telephone system was to be managed privately as a common carrier that could not arbitrarily deny service to anyone, and would be permitted to cross-subsidize local phone service to keep it affordable. The result has been the arrival of telephones in 94 percent of U.S. households.

Today the successors to the printing press remain as essential as ever in helping distributed groups of people respond to the excesses of central authority, whether by photocopied samizdat in Russia or faxes to Tiananmen Square demonstrators. Such communication enables a kind of political parallel processing that is a very clear threat to efforts to keep political programming centralized. National attempts to limit and regulate Internet access are increasingly doomed, as the means to send and receive bits gets simpler and the available channels proliferate. Borders are just too permeable to bits. While China was busy worrying about whether to allow official Internet access, packets were already being sent through a satellite link that had been set up for high-energy physics experiments.

Now constellations of low-earth-orbit satellites are being launched that will bring the convenience of a cell-phone network everywhere on the globe. Motorola's Iridium system comprises sixty-six such satellites. (It was originally named after the seventy-seventh element, iridium, because it was going to have seventy-seven satellites, but when it was reduced to sixty-six satellites the name wasn't changed because the sixty-sixth element is dysprosium.) Governments, and private companies, are launching spy satellites for commercial applications. These satellites can almost read a license plate from space. Soon, the only way to cut a country off from the rest of the world will be to build a dome over it. The free exchange of information makes abuses harder to cover up, and knowledge of alternatives harder to suppress.

We should now worry less about control *of* the means of communication and more about control *by* the means of communication. While we've been diligently protecting each new medium from manipulation by latter-day Leos, communications and computing have been merging so that the medium can not only become the message, it can make sure that you know it.

A ringing telephone once held the promise that someone was interested enough in you in particular to look up your number and call you. The phone numbers themselves were deemed to be public information, but the telephone companies kept careful control over the databases used to produce the phone books. This was a stable arrangement until the advent of cheap personal computers and massive storage on inexpensive compact disks. It was perfectly legal to ship a PC and a carton of phone books off to a low-wage country, where the information was laboriously typed in (a few times, actually, to catch mistakes) and sent back on a CD. Access to your telephone had been constrained by the difficulty of any one person using more than a few phone books; now it became available to anyone with a CD drive.

Phone numbers were just the beginning. Every time you call an 800 number, or use a credit card, you deposit a few bits of personal information in a computer somewhere. Where you called from, what you bought, how much you spent. Any one of these records is harmless enough, but taken together they paint a very personal picture of you. This is what Lotus and Equifax did in 1990, when they announced the Marketplace:Households product. By assembling legally available consumer information, they pieced together a database of 120 million Americans, giving names, addresses, ages, marital status, spending habits, estimated incomes. Interested in finding rich widows living on isolated streets? How about chronic credit-card overspenders? No problem. This shocking invasion of privacy was perfectly legal; the only thing that prevented the database from being sold was the tens of thousands of outraged e-mail complaints that Lotus received. Lotus's blunder only helped to pave the way for the slightly less indiscreet direct marketers that followed, cheerfully collecting and selling your particulars. These are the data that turn a ringing telephone from an invitation to personal communication to an announcement that you've been selected as a target demographic.

My wife and I spent a few years getting to know each other before we moved in together, making sure that we were compatible. I wasn't nearly so choosy about my telephone, although I certainly wouldn't tolerate from a spouse many of the things it does. The phone summons me when I'm in the shower and can't answer it, and when I'm asleep and don't want to answer it; it preserves universal access to me for friend and telemarketing foe alike. Letting the phone off the hook because it has no choice in whether to ring or not is akin to the military excuse that it's not responsible for its actions because it's only following orders. Bad people won't go away, but bad telephones can. A telephone that can't make these distinctions is not fit for polite company.

If a computer is connected to the telephone it's probably used for e-mail, and if it's used for e-mail there's probably too much of it. I get about one hundred messages a day. A surprising amount of that comes in the middle of the night, sent by a growing population of nocturnal zombies who arise from their beds to answer their e-mail before the next day's installment arrives. The computer's printer will need paper, and more paper, and more paper. In 1980 the United States consumed 16 million tons of paper for printing and writing; ten years later that jumped to 25 millions tons. So much for the paperless society. If you want to carry the computer around with you it needs batteries; the United States is currently throwing away 2 billion of them a year, the largest source of heavy metals in landfills.

There's a very real sense in which the things around us are infringing a new kind of right that has not needed protection until now. We're spending more and more time responding to the demands of machines. While relating religious oppression to poorly designed consumer electronics might appear to trivialize the former and selfishly complain about the latter, recognize that regimes that imposed these kinds of practices on their subjects have generally been overthrown at the earliest opportunity.

The first step in protecting rights is to articulate them. In keeping with the deflation from Luther's Ninety-five Theses to the Bill of Rights' ten amendments, I'd like to propose just three new ones:

BILL OF THINGS USERS' RIGHTS

You have the right to:

- Have information available when you want it, where you want it, and in the form you want it
- Be protected from sending or receiving information that you don't want
- Use technology without attending to its needs

A shoe computer that can get energy from walking instead of needing batteries, that can communicate through a body instead of requiring adapter cables, that can deliver a message to eyeglasses or a shirtsleeve or an earring, that can implement a cryptographic protocol to unobtrusively authorize a purchase without providing any identifying information, that can figure out when to speak and when not to, is not really an unusually good idea; it's just that a laptop that cannot do these things is an increasingly bad one.

By shifting more authority to people, the Reformation led to a new morality, a new set of shared standards of rights and responsibilities that helped define what it means to be civilized. The opportunities and excesses associated with the digital world now require that this morality be updated from the sixteenth century to the twenty-first. Oppressive machines are as bad as oppressive churches; freedom of technological expression is as important as freedom of religious expression.

As market forces drive down the cost of computing and communicating so that it can become a discretionary purchase for most people, like buying a TV, universal access is becoming less of a concern. A new barrier enforcing social stratification is ease of access. The members of my family have no trouble earning advanced degrees, yet they struggle to connect a computer to the network, and manage to find countless ways to lose files they are working on. They rely on a complex social and technical support network to solve these problems (me). If a Ph.D. is not sufficient qualification to use a computer, how can we hope that putting computers in the hands of more people can help them? Democratizing access to solutions, rather than technology, is going to require as much concern for the usability as the availability of the technology.

Our legal system is already straining to cope with the unexpected implications of connecting smart distributed systems. Establishing technological rights cannot happen by central command; it must happen by changing the expectations of both the designers and users of the new technology. Here, too, the first step is to articulate the basic requirements needed to meet our demands. With the current division of labor neither people nor things can do what they do best. Accordingly, I would also like to propose three rights for things.

BILL OF THINGS' RIGHTS

Things have the right to:

- Have an identity
- Access other objects
- Detect the nature of their environment

My office has ten things that include a clock, and each one reports a different time with equal confidence. I perform a high-tech equivalent of marking the solstices by the ritual of navigating through ten sets of menus to update the time. My annoyance at having to do this is pointless if these competing clocks are not given the resources needed to do what I expect of them, which is to tell me the time. That requires they know something about timekeeping, including determining where they are to set the time zone, communicating with a time standard to get the correct time, and even recognizing that most of them are redundant and don't all need to report the same information.

Taken together, these rights define a new notion of behavior, shared between people and machines, that is appropriate for a new era. Along with these rights come new responsibilities; what was suitable in the sixteenth century is not now. E-mail is a good example of how social norms suitable for the physical world can outlive their usefulness in the digital world.

My mother taught me to speak when spoken to. This reasonable instruction means that each day I should send out one hundred e-mail messages in response to the hundred or so that I receive from other people. Of these, about 10 percent need further action that requires initiating a new message to someone else. At the beginning of the day I had one hundred messages coming in; at the end I have 110 going out. If, then, each of my recipients is equally well-behaved, the next day there will be 121 circulating because of me. If each recipient remains as courteous, in a week my day's investment in correspondence will have paid a dividend of 214 messages. This exponential explosion will continue unabated until either people or networks break down. Of course some messages naturally don't need a response, but every day in which you send out more e-mail than you receive, you're responsible for contributing to e-mail inflation.

The problem stems from an asymmetry between the time it takes to create and read a message. While most people don't write any faster than they used to, it takes just moments to paste big stretches of other texts into a message, or forward messages on to others, or add many people as recipients to one message. What's not immediately apparent is the cost in other people's time. If I take a minute to skim each message I get and then a minute to dash off a hasty response, I've used up half a working day.

Given this mismatch, the most considerate thing to do is answer e-mail only if it can't be avoided, and to do so as briefly as possible. When I arrived in the Media Lab I used to write carefully crafted e-mail messages that addressed all sides of an issue, patiently, at length, taking as much space as needed to make sure that I said everything just right, not missing anything that warranted comment, or explanation, or observation. My peers' terse or absent responses left me wondering about their manners, if not their literacy, until each hour I was adding per day to do e-mail left me wondering about my sanity. In an era of overcommunication, saying less is more. This is entirely unlike what is appropriate for hand-written correspondence, where the effort to send a message exceeds the effort to receive one.

The spread of computing is making it ever easier to communicate anywhere, anytime; the challenge now is to make it easier to not communicate. We've come too far in connecting the world to be able to switch everything off; we must go further in turning everything on. Our devices must become smart enough to help us manage as well as move information.

One of the inspirations for the prevailing windows-and-mice user interface was Jean Piaget's studies of child development. Windows and mice represent a stage when infants begin to gesture to identify things in their environment. This was never meant to last as long as it has; like infants, interfaces should also grow up. Around the time that infants start pointing they also start talking. They practice language by incessant chatter, saying things over and over again for the pleasure of hearing themselves speak, but not yet understanding the repercussions of their actions. That's what so much electronic communication is, growing pains as our social expectations catch up to the technological means with which we live.

Amid this din, a quiet voice paradoxically cuts through with perhaps the best insight of all into communication in an Information Age. The original media hacker, Luther, somewhat grudgingly published a complete edition of his Latin works in 1545 to correct the errors that had accumulated in earlier copies. In the preface, he explains his hesitation to release one more contribution to his society's information overload: "I wanted all my books to be buried in perpetual oblivion, that thus there might be room for better books."

Bad Words

A group from a major computer company once came to the Media Lab to learn about our current research. The day started with their explaining that they wanted to see agent technology. Agents are a fashionable kind of computer program designed to learn users' preferences and help anticipate and act on their needs. Throughout the day the visitors saw many relevant things, including software to assist with collaboration among groups of people, environments for describing how programs can adapt, and mathematical techniques for finding patterns in user data. I almost fell out of my chair at the wrap-up at the end of the day when they asked when they were going to see agent technology. They had clearly been tasked to acquire some agents, but wouldn't recognize one if it bit them.

I spent a day consulting with a firm that spends billions of dollars a year on information technology. After working with them for most of the day on methods for analyzing and visualizing large data sets to find hidden structure, a senior executive asked if we could change subjects and talk about "data mining." He had heard about this great new technique that separates the nuggets of insight from the chaff of data that most companies have. Unfortunately, he had no idea that data mining and data analysis had anything to do with each other, and that in fact that was what we had been doing all day long.

At an industry lunch I found myself seated next to an editor for a major computer magazine. She was writing an article about exciting recent breakthroughs in artificial intelligence, and asked me if I used "neural net technology." When I told her that I did use nonlinear layered models to approximate functions with many variables, she looked concerned but gave me a second chance and asked if I used "fuzzy logic technology." After I answered that I did include probabilities in my models to let me account for advance beliefs and subsequent observations, she rolled her eyes and made clear what a burden it was to talk to someone who was so technologically unhip. She was particularly disappointed to see such old-fashioned thinking in the Media Lab.

Any software catalog is full of examples of perfectly unremarkable programs with remarkable descriptions. You can get ordinary games, or games that take advantage of virtual reality technology; regular image compression programs, or special ones that use fractal technology; plain word processors, or new and improved ones with artificial intelligence technology. The mysterious thing is that, other than the descriptions, the capabilities of these new programs are surprisingly similar to those of the old-fashioned ones.

These unfortunate examples share a kind of digitally enhanced semiotic confusion. I'm always suspicious when the word "technology" is injudiciously attached to perfectly good concepts that don't need it. Adding "technology" confers the authority of, say, a great Victorian steam engine on what might otherwise be a perfectly unremarkable computer program.

If you're offered a better mousetrap, you can inspect it and evaluate its design improvements. But if you're offered a better computer program, it is difficult to peer inside and judge how it works. Consequently, the words that are used to describe software can have more influence and less meaning than those used to describe a mousetrap. Because so many people are so unclear about where hardware leaves off and software begins, reasonable people accept unreasonable claims for computer programs. A market persists for software that purports to speed up the processor in a computer, or add memory to a system, even though those attributes are clearly questions of physical resources. It doesn't require an advanced testing laboratory to recognize that such things can't (and don't) work.

These problems are not just a consequence of malicious marketing—even the developers can be misled by their own labels. When I was a graduate student at Cornell doing experiments in the basement of the Physics building, there was a clear sign when someone's thesis research was in trouble: they made overly nice cases for their instruments. Creating and understanding new capabilities was difficult; creating fancy packages for old capabilities was labor-intensive but guaranteed to succeed. Making an elaborate housing provided the illusion of progress without the risk of failure because nothing new was actually being done. Something similar is happening with bad programs wrapped in elaborate interfaces and descriptions.

There are a number of words that I've found to be good indicators of such suspect efforts because they are used so routinely in ways that are innocently or intentionally misleading. Lurking behind terms such as "multimedia," "virtual reality," "chaos theory," "agents," "neural networks," and "fuzzy logic" are powerful ideas with fascinating histories and remarkable implications, but their typical use is something less than remarkable. Realization of their promise requires much more than the casual application of a label.

When Yo-Yo was playing our cello you could correctly call what he was doing "interactive multimedia" because a computer was making sounds in response to his inputs, and you might even call it a kind of "virtual reality" because he was using physical sensors to manipulate a virtual instrument. But you probably wouldn't do either because there was a much more compelling reality: Yo-Yo Ma playing the cello. This is another clue that one should worry about the use of these kinds of technological buzz phrases. A marketing description of the feature list of a Stradivarius could not distinguish it from a lesser instrument, or even from a multimedia PC running a cello simulator. In fact, the PC would come out much better in a list of standard features and available options and upgrades. The Strad, after all, can only make music. The words we usually use to describe computers have a hard time capturing the profound difference between things that work, and things that work very well. A Strad differs from a training violin in lots of apparently inconsequential details that when taken together make all of the difference in the world.

People are often surprised to find that there's no one in the Media Lab who would say that they're working on multimedia. When the lab started fifteen years ago, suggesting that computers should be able to speak and listen and see was a radical thing to do, so much so that it was almost grounds for being thrown off campus for not being serious. Now not only are there conferences and journals on multimedia, the best work being done has graduated from academic labs and can be found at your local software store. It was a battle for the last decade to argue that a digital representation lets computers be equally adept at handling text, audio, or video. That's been won; what matters now is what is done with these capabilities.

Studying multimedia in a place like the Media Lab makes as much sense as studying typewriters in a writers' colony, or electricity in a Computer Science department. Identifying with the tools deflects attention away from the applications that should motivate and justify them, and from the underlying skills that are needed to create them. The causal ease of invoking "multimedia" to justify any effort that involves sounds or images and computers is one of the reasons that there has been so much bad done in the name of multimedia.

When I fly I've found a new reason to look for airsickness bags: the inflight videos touting new technology. These hyperactive programs manage to simulate turbulence without even needing to leave the ground. They are the video descendents of early desktop publishing. When it became possible to put ten fonts on a page, people did, making so much visual noise. Good designers spend years thinking about how to convey visual information in ways that guide and satisfy the eye, that make it easy to find desired information and progress from a big picture to details. It requires no skill to fill a page with typography; it requires great discipline to put just enough of the right kind of information in the right place to communicate the desired message rather than a printer demo. In the same way, adding audio and video together is easy when you have the right software and hardware, but to be done well requires the wisdom of a film director, editor, cinematographer, and composer combined. Jump cuts look just as bad done with bits or atoms.

Overuse of the possibilities afforded by improving technology has come full circle in a style of visual design that miraculously manages to capture in print the essence of bad multimedia. It hides the content behind a layout of such mind-boggling complexity that, after I admire the designer's knowledge of the features of their graphics programs, I put the magazine down and pick up a nice book. Such cluttered design is certainly not conducive to distinguishing between what exists and what does not, between what is possible and what is not. And so here again in this kind of writing the labels attached to things take on more importance than they should.

After media became multi, reality became virtual. Displays were put into goggles and sensor gloves were strapped onto hands, allowing the wearer to move through computer-generated worlds. The problem with virtual reality is that it is usually taken to mean a place entirely disconnected from our world. Either you are in a physical reality, or a virtual one. When most people first actually encounter virtual reality, they're disappointed because it's not that unlike familiar video games. Although in part this is a consequence of describing simple systems with fancy terms, the analogy with video games goes deeper. In a virtual world there is an environment that responds to your actions, and the same is true of a video game. Virtual reality is no more or less than computers with good sensors and displays running models that can react in real time. That kind of capability is becoming so routine that it doesn't require a special name.

One of the defining features of early virtual reality was that it was completely immersive; the goal was to make video and audio displays that completely plugged your senses. That's become less important, because bad virtual reality has reminded people of all of the good features of the world of atoms. Instead of a sharp distinction between the virtual and the physical world, researchers are beginning to merge the best attributes of both by embedding the displays into glasses or walls. Discussions about virtual reality lead to awkward constructions like "real reality" to describe that which is not virtual; it's much more natural to simply think about reality as something that is presented to you by information in your environment, both logical and physical.

"Chaos theory" is a leading contender for a new paradigm to describe the complexity of that physical environment and bring it into the precise world of computers. I read a recent newspaper article that described the significance of ". . . chaos theory, which studies the disorder of formless matter and infinite space." Wow. One can't ask for much more than that. What makes "chaos theory" particularly stand out is that it is generally considered to be a theory only by those who don't work on it.

The modern study of chaos arguably grew out of Ed Lorenz's striking discovery at MIT in the 1960s of equations that have solutions that appear to be random. He was using the newly available computers with graphical displays to study the weather. The equations that govern it are much too complex to be solved exactly, so he had the computer find an approximate solution to a simplified model of the motion of the atmosphere. When he plotted the results he thought that he had made a mistake, because the graphs looked like random scribbling. He didn't believe that his equations could be responsible for such disorder. But, hard as he tried, he couldn't make the results go away. He eventually concluded that the solution was correct; the problem was with his expectations. He had found that apparently innocuous equations can contain solutions of unimaginable complexity. This raised the striking possibility that weather forecasts are so bad because it's fundamentally not possible to predict the weather, rather than because the forecasters are not clever enough.

Like all good scientific discoveries, the seeds of Lorenz's observation can be found much earlier. Around 1600 Johannes Kepler (a devout Lutheran) was trying to explain the observations of the orbits of the planets. His first attempt, inspired by Copernicus, matched them to the diameters of nested regular polyhedra (a pyramid inside a cube . . .). He published this analysis in the *Mysterium Cosmographicum,* an elegant and entirely incorrect little book. While he got the explanation wrong, this book did have all of the elements of modern science practice, including a serious comparison between theoretical predictions and experimental observations, and a discussion of the measurement errors. Armed with this experience plus better data he got it right a few years later, publishing a set of three laws that could correctly predict the orbits of the planets. The great triumph of Newton's theory of gravitation around 1680 was to derive these laws as consequences of a gravitational force acting between the planets and the sun. Newton tried to extend the solution to three bodies (such as the combined earth-moon-sun system), but failed and concluded that the problem might be too hard to solve. This was a matter of more than passing curiosity because it was not known if the solution of the three-body problem was stable, and hence if there was a chance of the earth spiraling into the sun.

Little progress was made on the problem for many years, until around 1890 the French mathematician Henri Poincaré was able to prove that Newton's hunch was right. Poincaré showed that it was not possible to write down a simple solution to the three-body problem. Lacking computers, he could only suspect what Lorenz was later able to see: the solutions could not be written down because their behavior over time was so complex. He also realized that this complexity would cause the solutions to depend sensitively on any small changes. A tiny nudge to one planet would cause all of the trajectories to completely change.

This behavior is familiar in an unstable system, such as a pencil that is balanced on its point and could fall in any direction. The converse is a stable system, such as the position of the pendulum in a grandfather clock that hangs vertically if the clock is not wound. Poincaré encountered, and Lorenz developed, the insight that both behaviors could effectively occur at the same time. The atmosphere is not in danger of falling over like a pencil, but the flutter of the smallest butterfly can get magnified to eventually change the weather pattern. This balance between divergence and convergence we now call chaos.

Ordinary mathematical techniques fail on chaotic systems, which appear to be random but are governed by simple rules. The discovery of chaos held out the promise that simple explanations might be lurking behind nature's apparent complexity. Otherwise eminent scientists relaxed their usual critical faculties to embrace the idea. The lure of this possibility led to the development of new methods to recognize and analyze chaos. The most remarkable of these could take almost any measured signal, such as the sound of a faucet dripping, and reveal the behavior of the unseen parts of the system producing the signal. When applied to a dripping faucet this technique did in fact provide a beautiful explanation for the patterns in what had seemed to be just an annoying consequence of leaky washers. After this result, the hunt was on to search for chaos.

Lo and behold, people found it everywhere. It was in the stars, in the weather, in the oceans, in the body, in the markets. Computer programs were used to test for chaos by counting how many variables were needed to describe the observations; a complex signal that could be explained by a small number of variables was the hallmark of chaos. Unfortunately, this method had one annoying feature. When applied to a data set that was too small or too noisy it could erroneously conclude that anything was chaotic. This led to excesses such as the *Economist* magazine reporting that "Mr. Peters thinks that the S&P 500 index has 2.33 fractal dimensions." This means that future values of the stock market could be predicted given just three previous values, a recipe for instant wealth if it wasn't so obviously impossible. Such nonsensical conclusions were accepted on the basis of the apparent authority of these computer programs.

Chaos has come to be associated with the study of anything complex, but in fact the mathematical techniques are directly applicable only to simple systems that appear to be complex. There has proved to be a thin layer between systems that appear to be simple and really are, and those that appear to be complex and really are. The people who work on chaos are separating into two groups, one that studies the exquisite structure in the narrow class of systems where it does apply, and another that looks to use the methods developed from the study of chaos to help understand everything else. This leaves behind the frequently noisy believers in "chaos theory," inspired but misled by the exciting labels.

They're matched in enthusiasm by the "agents" camp, proponents of computer programs that learn user preferences and with some autonomy act on their behalf. The comforting images associated with an agent are a traditional English butler, or a favorite pet dog. Both learn to respond to their master's wishes, even if the master is not aware of expressing them. Nothing would be more satisfying than a digital butler that could fetch. Unfortunately, good help is as hard to find in the digital world as it is in the physical one.

Agents must have a good agent. The widespread coverage of them has done a great job of articulating the vision of what an agent *should* be able to do, but it's been less good at covering the reality of what agents *can* do. Whatever you call it, an agent is still a computer program. To write a good agent program you need to have reasonable solutions to the interpretation of written or spoken language and perhaps video recognition so that it can understand its instructions, routines for searching through large amounts of data to find the relevant pieces of information, cryptographic schemes to manage access to personal information, protocols that allow commerce among agents and the traditional economy, and computer graphics techniques to help make the results intelligible. These are hard problems. Wanting to solve them is admirable, but naming the solution is not the same as obtaining it.

Unlike what my confused industrial visitors thought, an agent is very much part of this world of algorithms and programming rather than a superhuman visitor from a new world. The most successful agents to date have bypassed many of these issues by leveraging human intelligence to mediate interactions that would not happen otherwise. Relatively simple programs can look for people who express similar preferences in some areas such as books or recordings, and then make recommendations based on their previous choices. That's a great thing to do, helping realize the promise of the Internet to build communities and connections among people instead of isolating them, but then it becomes as much a question of sociology as programming to understand how, where, and why people respond to each other's choices.

If an agent is ever to reason with the wisdom of a good butler, it's natural to look to the butler's brain for insight into how to do this. Compared to a computer, the brain is made out of slow, imperfect components, yet it is remarkably powerful, reliable, and efficient. Unlike a conventional digital computer, it can use continuous analog values, and it takes advantage of an enormous number of simple processing elements working in parallel, the neurons. These are "programmed" by varying the strength of the synaptic connections among them. "Neural networks" are mathematical models inspired by the success of this architecture.

In the 1940s mathematical descriptions were developed for neurons and their connections, suggesting that it might be possible to go further to understand how networks of neurons function. This agenda encountered an apparently insurmountable obstacle in 1969 when Marvin Minsky and Seymour Papert proved that a layer of neurons can implement only the simplest of functions between their inputs and outputs. The strength of their result effectively halted progress until the 1980s when a loophole was found by introducing neurons that are connected only to other neurons, not inputs or outputs. With such a hidden layer it was shown that a network of neurons could represent any function.

Mathematical models traditionally have been based on finding the values of adjustable parameters to most closely match a set of observations. If a hidden layer is used in a neural network, this is no longer possible. It can be shown that there's no feasible way to choose the best values for the connection strengths. This is analogous to the study of chaos, where the very complexity that makes the equations impossible to solve makes them valuable to use. The behavior of chaotic equations can be understood and applied even if it can't be exactly predicted. Similarly, it turned out that hidden layers can be used by searching for reasonable weights without trying to find the best ones. Because the networks are so flexible, even a less-than-ideal solution can be far more useful than the exact solution of a less-capable model. As a consequence of this property, neural networks using surprisingly simple search strategies were surprisingly capable.

The combination of some early successes and language that suggests that neural networks work the same way the brain does led to the misleading impression that the problem of making machines think had been solved. People still have to think to use a neural network. The power, and problems, of neural networks were amply demonstrated in a study that I ran at the Santa Fe Institute. It started with a workshop that I attended there, exploring new mathematical techniques for modeling complex systems. The meeting was distressingly anecdotal, full of sweeping claims for new methods but containing little in the way of insight into how they fail or how they are related to what is already known. In exasperation I made a joke and suggested that we should have a data analysis contest. No one laughed, and in short order the Santa Fe Institute and NATO had agreed to support it.

My colleague Andreas Weigend and I selected interesting data sets from many disciplines, giving the changes over time in a currency exchange rate, the brightness of a star, the rhythm of a heartbeat, and so forth. For extra credit we threw in the end of *The Art of the Fugue,* the incomplete piece that Bach was writing when he died. These data were distributed around the world through the Internet. Researchers were given quantitative questions appropriate to each domain, such as forecasting future values of the series. These provided a way to make comparisons across disciplines independent of the language used to describe any particular technique.

From the responses I learned as much about the sociology of science as I did about the content. Some people told us that our study was a mistake because science is already too competitive and it's a bad influence to try to make these comparisons; others told us that our study was doomed because it's impossible to make these kinds of comparisons. Both were saying that their work was not falsifiable. Still others said that our study was the best thing that had happened to the field, because they wanted to end the ambiguity of data analysis and find out which technique was the winner.

One of the problems presented data from a laser that was erratically fluctuating on and off; the task was to predict how this pattern would continue after the end of the data set. Because the laser was chaotic, traditional forecasting methods would not do much better than guessing randomly. Some of the entires that we received were astoundingly good. One of the best used a neural network to forecast the series, and it was convincingly able to predict all of the new behavior that it had not seen in the training data. Here was a compelling demonstration of the power of a neural net.

For comparison, one entry was done by eye, simply guessing what would come next. Not surprisingly, this one did much worse than the neural network. What was a surprise was that it beat some of other entries by just as large a margin. One team spent hours of supercomputer time developing another neural network model; it performed significantly worse than the visual inspection that took just a few moments. The best and the worse neural networks had similar architectures. Nothing about their descriptions would indicate the enormous difference in their performance; that was a consequence of the insight with which the networks were applied.

It should not be too unexpected that apparently similar neural networks can behave so differently—the same is true of real brains. As remarkable as human cognition can be, some people are more insightful than others. Starting with the same hardware, they differ in how they use it. Using a neural network gives machines the same opportunity to make mistakes that people have always enjoyed.

Many experts in neural networks don't even study neural networks anymore. Neural nets provided the early lessons that started them down the path of using flexible models that learn, but overcoming the liabilities of neural nets has led them to leave behind any presumption of modeling the brain and focus directly on understanding the mathematics of reasoning. The essence of this lies in finding better ways to manage the tension between experience and beliefs.

One new technique on offer to do that is "fuzzy logic." It is sold as an entirely new kind of reasoning that replaces the sharp binary decisions forced by our Western style of thinking with a more Eastern sense of shades of meaning that can better handle the ambiguity of the real world. In defense of the West, we've known for a few centuries how to use probability theory to represent uncertainty. If you force a fuzzy logician to write down the expressions they use, instead of telling you the words they attach to them, it turns out that the expressions are familiar ones with new names for the terms. That itself is not so bad, but what is bad is that such naming deflects attention away from the much better developed study of probability theory.

This danger was on display at a conference I attended on the mathematics of inference. A battle was raging there between the fuzzy logic camp and the old-fashioned probabilists. After repeatedly pushing the fuzzy logicians to show anything at all that they could do that could not be done with regular probability theory, the fuzzy side pulled out their trump card and told the story of a Japanese helicopter controller that didn't work until fuzzy logic was used. Everyone went home feeling that they had won the argument. Of course a demonstration that something does work is far from proof that it works better than, or even differently than, anything else. What's unfortunate about this example is that perfectly intelligent people were swayed by the label attached to the program used in the helicopter rather than doing the homework needed to understand how it works and how it relates to what is already known. If they had done so, they might have discovered that the nonfuzzy world has gone on learning ever more interesting and useful things about uncertainty.

Connecting all of these examples is a belief in magic software bullets, bits of code that can solve the hard problems that had stumped the experts who didn't know about neural networks, or chaos, or agents. It's all too easy to defer thinking to a seductive computer program. This happens on the biggest scales. At still one more conference on mathematical modeling I sat through a long presentation by someone from the Defense Department on how they are spending billions of dollars a year on developing mathematical models to help them fight wars. He described an elaborate taxonomy of models of models of models. Puzzled, at the end of it I hazarded to put up my hand and ask a question that I thought would show everyone in the room that I had slept through part of the talk (which I had). I wondered whether he had any idea whether his billion-dollar models worked, since it's not convenient to fight world wars to test them. His answer, roughly translated, was to shrug and say that that's such a hard question they don't worry about it. Meanwhile, the mathematicians in the former Soviet Union labored with limited access to computers and had no recourse but to think. As a result a surprising fraction of modern mathematical theory came from the Soviet Union, far out of proportion to its other technical contributions.

Where once we saw farther by standing on the shoulders of our predecessors, in far too many cases we now see less by standing on our predecessors' toes. You can't be interdisciplinary without the disciplines, and without discipline. Each of the problematical terms I've discussed is associated with a very good idea. In the life cycle of an idea, there is a time to tenderly nurse the underlying spark of new insight, and a time for it to grow up and face hard questions about how it relates to what is already known, how it generalizes, and how it can be used. Even if there aren't good answers, these ideas can still be valuable for how they influence people to work on problems that do lead to answers. The Information Age is now of an age that deserves the same kind of healthy skepticism applied to the world of bits that we routinely expect in the world of atoms.

Bit Beliefs

For as long as people have been making machines, they have been trying to make them intelligent. This generally unsuccessful effort has had more of an impact on our own ideas about intelligence and our place in the world than on the machines' ability to reason. The few real successes have come about either by cheating, or by appearing to. In fact, the profound consequences of the most mundane approaches to making machines smart point to the most important lesson that we must learn for them to be able to learn: intelligence is connected with experience. We need all of our senses to make sense of the world, and so do computers.

A notorious precedent was set in Vienna in 1770, when Wolfgang van Kempelen constructed an automaton for the amusement of Empress Maria Theresa. This machine had a full-size, mustachioed, turbaned Turk seated before a chess board, and an ingenious assortment of gears and pulleys that enabled it to move the chess pieces. Incredibly, after van Kempelen opened the mechanism up for the inspection and admiration of his audiences and then wound it up, it could play a very strong game of chess. His marvel toured throughout Europe, amazing Benjamin Franklin and Napoleon.

After von Kempelen's death the Turk was sold in 1805 to Johann Maelzel, court mechanic for the Habsburgs, supporter of Beethoven, and inventor of the metronome. Maelzel took the machine further afield, bringing it to the United States. It was there in 1836 that a budding newspaper reporter, Edgar Allan Poe, wrote an exposé duplicating an earlier analysis in London explaining how it worked. The base of the machine was larger than it appeared; there was room for a small (but very able) chess player to squeeze in and operate the machine.

While the Turk might have been a fake, the motivation behind it was genuine. A more credible attempt to build an intelligent machine was made by Charles Babbage, the Lucasian Professor of Mathematics at Cambridge from 1828 to 1839. This is the seat that was held by Sir Isaac Newton, and is now occupied by Stephen Hawking. Just in case there was any doubt about his credentials, his full title was "Charles Babbage, Esq., M.A., F.R.S., F.R.S.E., F.R.A.S., F. Stat. S., Hon. M.R.I.A., M.C.P.S., Commander of the Italian Order of St. Maurice and St. Lazarus, Inst. Imp. (Acad. Moral.) Paris Corr., Acad. Amer. Art. et Sc. Boston, Reg. Oecon. Boruss., Phys. Hist. Nat. Genev., Acad. Reg. Monac., Hafn., Massil., et Divion., Socius. Acad. Imp. et Reg. Petrop., Neap., Brux., Patav., Georg. Floren., Lyncei. Rom., Mut., Philomath. Paris, Soc. Corr., etc." No hidden compartments for him.

Babbage set out to make the first digital computer, inspired by the Jacquard looms of his day. The patterns woven into fabrics by these giant machines were programmed by holes punched in cards that were fed to them. Babbage realized that the instructions could just as well represent the sequence of instructions needed to perform a mathematical calculation. His first machine was the Difference Engine, intended to evaluate quantities such as the trigonometric functions used by mariners to interpret their sextant readings. At the time these were laboriously calculated by hand and collected into error-prone tables. His machine mechanically implemented all of the operations that we take for granted in a computer today, reading in input instructions on punched cards, storing variables in the positions of wheels, performing logical operations with gears, and delivering the results on output dials and cards. Because the mechanism was based on discrete states, some errors in its operation could be tolerated and corrected. This is why we still use digital computers today.

Babbage oversaw the construction of a small version of the Difference Engine before the project collapsed due to management problems, lack of funding, and the difficulty of fabricating such complex mechanisms to the required tolerances. But these mundane details didn't stop him from turning to an even more ambitious project, the Analytical Engine. This was to be a machine that could reason with abstract concepts and not just numbers. Babbage and his accomplice, Lady Ada Lovelace, realized that an engine could just as well manipulate the symbols of a mathematical formula. Its mechanism could embody the rules for, say, calculus and punch out the result of a derivation. As Lady Lovelace put it, "the Analytical Engine weaves algebraical patterns just as the Jacquard Loom weaves flowers and leaves."

Although Babbage's designs were correct, following them went well beyond the technological means of his day. But they had an enormous impact by demonstrating that a mechanical system could perform what appear to be intelligent operations. Darwin was most impressed by the complex behavior that Babbage's engines could display, helping steer him to the recognition that biological organization might have a mechanistic explanation. In Babbage's own memoirs, *Passages from the Life of a Philosopher,* he made the prescient observation that

> It is impossible to construct machinery occupying unlimited
> space; but it is possible to construct finite machinery, and to use
> it through unlimited time. It is this substitution of the infinity of

time for the infinity of space which I have made use of, to limit
the size of the engine and yet to retain its unlimited power.

The programmability of his engines would permit them, and their
later electronic brethren, to perform many different functions with
the same fixed mechanism.

As the first computer designer it is fitting that he was also the
first to underestimate the needs of the market, saying, "I propose in
the Engine I am constructing to have places for only a thousand
constants, because I think it will be more than sufficient." Most
every computer since has run into the limit that its users wanted to
add more memory than its designers thought they would ever need.
He was even the first programmer to complain about lack of stan-
dardization:

> I am unwilling to terminate this chapter without reference to
> another difficulty now arising, which is calculated to impede the
> progress of Analytical Science. The extension of analysis is so
> rapid, its domain so unlimited, and so many inquirers are en-
> tering into its fields, that a variety of new symbols have been in-
> troduced, formed on no common principles. Many of these are
> merely new ways of expressing well-known functions. Unless
> some philosophical principles are generally admitted as the ba-
> sis of all notation, there appears a great probability of intro-
> ducing the confusion of Babel into the most accurate of all
> languages.

Babbage's frustration was echoed by a major computer company
years later in a project that set philosophers to work on coming up
with a specification for the theory of knowledge representation, an
ontological standard, to solve the problem once and for all. This effort was as unsuccessful, and interesting, as Babbage's engines.

Babbage's notion of one computer being able to compute anything was picked up by the British mathematician Alan Turing. He was working on the "Entscheidungsproblem," one of a famous group of open mathematical questions posed by David Hilbert in 1900. This one, the tenth, asked whether a mathematical procedure could exist that could decide the validity of any other mathematical statement. Few questions have greater implications. If the answer is yes, then it could be possible to automate mathematics and have a machine prove everything that could ever be known. If not, then it would always be possible that still greater truths could lay undiscovered just beyond current knowledge.

In 1936 Turing proved the latter. To do this, he had to bring some kind of order to the notion of a smart machine. Since he couldn't anticipate all the kinds of machines people might build, he had to find a general way to describe their capabilities. He did this by introducing the concept of a Universal Turing Machine. This was a simple machine that had a tape (possibly infinitely long), and a head that could move along it, reading and writing marks based on what was already on the tape. Turing showed that this machine could perform any computation that could be done by any other machine, by preparing it first with a program giving the rules for interpreting the instructions for the other machine. With this result he could then prove or disprove the Entscheidungsproblem for his one machine and have it apply to all of the rest. He did this by showing that it was impossible for a program to exist that could determine whether another program would eventually halt or keep running forever.

Although a Turing machine was a theoretical construction, in the period after World War II a number of laboratories turned to successfully making electronic computers to replace the human "computers" who followed written instructions to carry out calculations. These machines prompted Turing to pose a more elusive question:

could a computer be intelligent? Just as he had to quantify the notion of a computer to answer Hilbert's problem, he had to quantify the concept of intelligence to even clearly pose his own question. In 1950 he connected the seemingly disparate worlds of human intelligence and digital computers through what he called the Imitation Game, and what everyone else has come to call the Turing test. This presents a person with two computer terminals. One is connected to another person, and the other to a computer. By typing questions on both terminals, the challenge is to determine which is which. This is a quantitative test that can be run without having to answer deep questions about the meaning of intelligence.

Armed with a test for intelligence, Turing wondered how to go about developing a machine that might display it. In his elegant essay "Computing Machinery and Intelligence," he offers a suggestion for where to start:

> We may hope that machines will eventually compete with men
> in all purely intellectual fields. But which are the best ones to
> start with? Even this is a difficult decision. Many people think
> that a very abstract activity, like the playing of chess, would be
> best.

Turing thought so; in 1947 he was able to describe a chess-playing computer program. Since then computer chess has been studied by a who's who of computing pioneers who took it to be a defining challenge for what came to be known as Artificial Intelligence. It was thought that if a machine could win at chess it would have to draw on fundamental insights into how humans think. Claude Shannon, the inventor of Information Theory, which provides the foundation for modern digital communications, designed a simple chess program in 1949 and was able to get it running to play endgames. The first program that could play a full game of chess was developed at IBM in 1957, and an MIT computer won the first tournament match against a human player in 1967. The first grand-master lost a game to a computer in 1977.

A battle raged among computer chess developers between those who thought that it should be approached from the top down, studying how humans are able to reason so effectively with such slow processors (their brains), and those who thought that a bottom-up approach was preferable, simply throwing the fastest available hardware at the problem and checking as many moves as possible into the future. The latter approach was taken in 1985 by a group of graduate students at Carnegie Mellon who were playing hooky from their thesis research to construct a computer chess machine. They used a service just being made available through the Defense Advanced Research Projects Agency (DARPA) to let researchers design their own integrated circuits; DARPA would combine these into wafers that were fabricated by silicon foundries. The Carnegie Mellon machine was called "Deep Thought" after the not-quite-omniscient supercomputer in Douglas Adams's *Hitchhiker's Guide to the Galaxy.*

In 1988 the world chess champion Gary Kasparov said there was "no way" a grandmaster would be defeated by a computer in a tournament before 2000. Deep Thought did that just ten months later. IBM later hired the Carnegie Mellon team and put a blue suit on the machine, renaming it "Deep Blue." With a little more understanding of chess and a lot faster processors, Deep Blue was able to evaluate 200 million positions per second, letting it look fifteen to thirty moves ahead. Once it could see that far ahead its play took on occasionally spookily human characteristics. Deep Blue beat Kasparov in 1997.

Among the very people you would expect to be most excited about this realization of Turing's dream, the researchers in artificial intelligence, the reaction to the victory has been curiously muted. There's a sense that Deep Blue is little better than van Kempelen's Turk. Nothing was learned about human intelligence by putting a human inside a machine, and the argument holds that nothing has been learned by putting custom chips inside a machine. Deep Blue is seen as a kind of idiot savant, able to play a good game of chess without understanding why it does what it does.

This is a curious argument. It retroactively adds a clause to the Turing test, demanding that not only must a machine be able to match the performance of humans at quintessentially intelligent tasks such as chess or conversation, but the way that it does so must be deemed to be satisfactory. Implicit in this is a strong technological bias, favoring a theory of intelligence appropriate for a particular kind of machine. Although brains can do many things in parallel they do any one thing slowly; therefore human reasoning must use these parallel pathways to best advantage. Early computers were severely limited by speed and memory, and so useful algorithms had to be based on efficient insights into a problem. More recent computers relax these constraints so that a brute-force approach to a problem can become a viable solution. No one of these approaches is privileged—each can lead to a useful kind of intelligence in a way that is appropriate for the available means. There's nothing fundamental about the constraints associated with any one physical mechanism for manipulating information.

The question of machine intelligence is sure to be so controversial because it is so closely linked with the central mystery of human experience, our consciousness. If a machine behaves intelligently, do we have to ascribe a kind of awareness to it? If we do, then the machine holds deep lessons about the essence of our own experience; if not, it challenges the defining characteristic of being human. Because our self-awareness is simultaneously so familiar and so elusive, most every mechanism that we know of gets pressed into service in search of an explanation. One vocal school holds that quantum mechanics is needed to explain consciousness. Quantum mechanics describes how tiny particles behave. It is a bizarre world, remote from our sensory experience, in which things can be in many places at the same time, and looking at something changes it. As best I've been able to reconstruct this argument, the reasoning is that (1) consciousness is mysterious, (2) quantum mechanics is mysterious, (3) nonquantum attempts to explain consciousness have failed, therefore (4) consciousness is quantum-mechanical. This is a beautiful belief that is not motivated by any experimental evidence, and does not directly lead to testable experimental predictions. Beliefs about our existence that are not falsifiable have a central place in human experience—they're called religion.

I spent an intriguing and frustrating afternoon running over this argument with an eminent believer in quantum consciousness. He agreed that the hypothesis was not founded on either experimental evidence or testable predictions, and that beliefs that are matters of faith rather than observation are the domain of religion rather than science, but then insisted that his belief was a scientific one. This is the kind of preordained reasoning that drove Turing to develop his test in the first place; perhaps an addendum is needed after all to ask the machine how it feels about winning or losing the test.

The very power of the machines that we construct turns them into powerful metaphors for explaining the world. When computing was done by people rather than machines, the technology of reasoning was embodied in a pencil and a sheet of paper. Accordingly, the prevailing description of the world was matched to that representation. Newton and Leibniz's theory of calculus, developed around 1670, provided a notation for manipulating symbols representing the value and changes of continuous quantities such as the orbits of the planets. Later physical theories, like quantum mechanics, are based on this notation.

At the same time Leibniz also designed a machine for multiplying and dividing numbers, extending the capabilities of Pascal's 1645 adding machine. These machines used gears to represent numbers as discrete rather than continuous quantities because otherwise errors would inevitably creep into their calculations from mechanical imperfections. While a roller could slip a small amount, a gear slipping is a much more unlikely event. When Babbage started building machines to evaluate not just arithmetic but more complex functions he likewise used discrete values. This required approximating the continuous changes by small differences, hence the name of the Difference Engine. These approximations have been used ever since in electronic digital computers to allow them to manipulate models of the continuous world.

Starting in the 1940s with John von Neumann, people realized that this practice was needlessly circular. Most physical phenomena start out discrete at some level. A fluid is not actually continuous; it is just made up of so many molecules that it appears to be continuous. The equations of calculus for a fluid are themselves an approximation of the rules for how the molecules behave. Instead of approximating discrete molecules with continuous equations that then get approximated with discrete variables on a computer, it's possible go directly to a computer model that uses discrete values for time and space. Like the checkers on a checkerboard, tokens that each represent a collection of molecules get moved among sites based on how the neighboring sites are occupied.

This idea has come to be known as Cellular Automata (CAs). From the 1970s onward, the group of Ed Fredkin, Tomaso Toffoli, and Norm Margolus at MIT started to make special-purpose computers designed for CAs. Because these machines entirely dispense with approximations of continuous functions, they can be much simpler and faster. And because a Turing machine can be described this way, a CA can do anything that can be done with a conventional computer.

A cellular automata model of the universe is no less fundamental than one based on calculus. It's a much more natural description if a computer instead of a pencil is used to work with the model.

And the discretization solves another problem: a continuous quantity can represent an infinite amount of information. All of human knowledge could be stored in the position of a single dot on a piece of paper, where the exact location of the dot is specified to trillions of digits and the data is stored in the values of those digits. Of course a practical dot cannot be specified that precisely, and in fact we believe that the amount of information in the universe is finite so that there must be a limit. This is built into a CA from the outset.

Because of these desirable features, CAs have grown from a computational convenience into a way of life. For the true believers they provide an answer to the organization of the universe. We've found that planets are made of rocks, that rocks are made up of atoms, the atoms are made up of electrons and the nucleus, the nucleus is made up of protons and neutrons, which in turn are made up of quarks. I spent another puzzling afternoon with a CA guru, who was explaining to me that one needn't worry about this subdivision continuing indefinitely because it must ultimately end with a CA. He agreed that experiments might show behavior that appears continuous, but that just means that they haven't looked finely enough. In other words, his belief was not testable. In other words, it was a matter of personal faith. The architecture of a computer becomes a kind of digital deity that brings order to the rest of the world for him.

Like any religion, these kinds of beliefs are enormously important in guiding behavior, and like any religion, dogmatic adherence to them can obscure alternatives. I spent one more happily exasperating afternoon debating with a great cognitive scientist how we will recognize when Turing's test has been passed. Echoing Kasparov's "no way" statement, he argued that it would be a clear epochal event, and certainly is a long way off. He was annoyed at my suggestion that the true sign of success would be that we cease to find the test interesting, and that this is already happening. There's a practical sense in which a modern version of the Turing test is being passed on a daily basis, as a matter of some economic consequence.

A cyber guru once explained to me that the World Wide Web had no future because it was too hard to figure out what was out there. The solution to his problem has been to fight proliferation with processing. People quickly realized that machines instead of people could be programmed to browse the Web, collecting indices of everything they find to automatically construct searchable guides. These search engines multiplied because they were useful and lucrative. Their success meant that Web sites had to devote more and more time to answering the rapid-fire requests coming to them from machines instead of the target human audience. Some Web sites started adding filters to recognize the access patterns of search engines and then deny service to that address. This started an arms race. The search engines responded by programming in behavior patterns that were more like humans. To catch these, the Web sites needed to refine their tests for distinguishing between a human and machine. A small industry is springing up to emulate, and detect, human behavior.

Some sites took the opposite tack and tried to invite the search engines in to increase their visibility. The simplest way to do this was to put every conceivable search phrase on a Web page so that any query would hit that page. This led to perfectly innocent searches finding shamelessly pornographic sites that just happen to mention airline schedules or sports scores. Now it was the search engine's turn to test for human behavior, adding routines to test to see if the word use on a page reflects language or just a list. Splitting hairs, they had to further decide if the lists of words they did find reflected reasonable uses such as dictionaries, or just cynical Web bait.

Much as Gary Kasparov might feel that humans can still beat

computers at chess in a more fair tournament, or my colleague thinks that the Turing test is a matter for the distant future, these kinds of reasoning tasks are entering into an endgame. Most computers can now beat most people at chess; programming human behavior has now become a job description. The original goals set for making intelligent machines have been accomplished.

Still, smart as computers may have become, they're not yet wise. As Marvin Minsky points out, they lack the common sense of a six-year-old. That's not too surprising, since they also lack the life experience of a six-year-old. Although Marvin has been called the father of artificial intelligence, he feels that that pursuit has gotten stuck. It's not because the techniques for reasoning are inadequate; they're fine. The problem is that computers have access to too little information to guide their reasoning. A blind, deaf, and dumb computer, immobilized on a desktop, following rote instructions, has no chance of understanding its world.

The importance of perception to cognition can be seen in the wiring of our brains. Our senses are connected by two-way channels: information goes in both directions, letting the brain fine-tune how we see and hear and touch in order to learn the most about our environment.

This insight takes us back to Turing. He concludes "Computing Machinery and Intelligence" with an alternative suggestion for how to develop intelligent machines:

> It can also be maintained that it is best to provide the machine with the best sense organs that money can buy, and then teach it to understand and speak English. This process could follow the normal teaching of a child. Things would be pointed out and named, etc.

The money has been spent on the computer's mind instead. Perhaps it's now time to remember that they have bodies, too.

Seeing Through Windows

I vividly recall a particular car trip from my childhood because it was when I invented the laptop computer. I had seen early teletype terminals; on this trip I accidentally opened a book turned on its side and realized that there was room on the lower page for a small typewriter keyboard, and on the upper page for a small display screen. I didn't have a clue how to make such a thing, or what I would do with it, but I knew that I had to have one. I had earlier invented a new technique for untying shoes, by pulling on the ends of the laces; I was puzzled and suspicious when my parents claimed prior knowledge of my idea. It would take me many more years to discover that Alan Kay had anticipated my design for the laptop and at that time was really inventing the portable personal computer at Xerox's Palo Alto Research Center (PARC).

Despite the current practice of putting the best laptops in the hands of business executives rather than children, my early desire to use a laptop is much closer to Alan's reasons for creating one. Alan's project was indirectly inspired by the work of the Swiss psychologist Jean Piaget, who from the 1920s onward spent years and years studying children. He came to the conclusion that what adults see as undirected play is actually a very structured activity. Children work in a very real sense as little scientists, continually positing and testing theories for how the world works. Through their endless interactive experiments with the things around them, they learn first how the physical world works, and then how the world of ideas works. The crucial implication of Piaget's insight is that learning cannot be restricted to classroom hours, and cannot be encoded in lesson plans; it is a process that is enabled by children's interaction with their environment.

Seymour Papert, after working with Piaget, brought these ideas to MIT in the 1960s. He realized that the minicomputers just becoming available to researchers might provide the ultimate sandbox for children. While the rest of the world was developing programming languages for accountants and engineers, Seymour and his collaborators created LOGO for children. This was a language that let kids express abstract programming constructs in simple intuitive terms, and best of all it was interfaced to physical objects so that programs could move things outside of the computer as well as inside it. The first one was a robot "turtle" that could roll around under control of the computer, moving a pen to make drawings.

Infected by the meme of interactive technology for children, Alan Kay carried the idea to the West Coast, to Xerox's Palo Alto Research Center. In the 1970s, he sought to create what he called a Dynabook, a portable personal knowledge navigator shaped like a notebook, a fantasy amplifier. The result was most of the familiar elements of personal computing.

Unlike early programming languages that required a specification of a precise sequence of steps to be executed, modern object-oriented languages can express more complex relationships among abstract objects. The first object-oriented programming language was Smalltalk, invented by Alan to let children play as easily with symbolic worlds as they do with physical ones. He then added interface components that were being developed by Doug Engelbart up the road at the Stanford Research Institute.

Doug was a radar engineer in World War II. He realized that a computer could be more like a radar console than a typewriter, interactively drawing graphics, controlled by an assortment of knobs and levers. Picking up a theme that had been articulated by Vannevar Bush (the person most responsible for the government's support of scientific research during and after the war) in 1945 with his proposal for a mechanical extender of human memory called a Memex, Doug understood that such a machine could help people navigate through the increasingly overwhelming world of information. His colleagues thought that he was nuts.

Computers were specialized machines used for batch processing, not interactive personal appliances. Fortunately, Engelbart was able to attract enough funding to set up a laboratory around the heretical notion of studying how people and computers might better interact. These ideas had a coming out in a rather theatrical demo he staged in San Francisco in 1968, showing what we would now recognize as an interactive computer with a mouse and multiple windows on a screen.

In 1974 these elements came together in the Xerox Alto prototype, and reached the market in Xerox's Star. The enormous influence of this computer was matched by its enormous price tag, about $50,000. This was a personal computer that only big corporations could afford. Windows and mice finally became widely available and affordable in Apple's Macintosh, inspired by Steve Jobs's visit to PARC in 1979, and the rest of personal computing caught up in 1990 when Microsoft released Windows 3.0.

The prevailing paradigm for how people use computers hasn't really changed since Englebart's show in 1968. Computers have proliferated, their performance has improved, but we still organize information in windows and manipulate it with a mouse. For years the next big interface has been debated. There is a community that studies such things, called Human-Computer Interactions. To give you an idea of the low level of that discussion, one of the most thoughtful HCI researchers, Bill Buxton (chief scientist at Silicon Graphics), is known for the insight that people have two hands. A mouse forces you to manipulate things with one hand alone; Bill develops interfaces that can use both hands.

A perennial contender on the short list for the next big interface is speech recognition, promising to let us talk to our computers as naturally as we talk to each other. Appealing as that is, it has a few serious problems. It would be tiring if we had to spend the day speaking continuously to get anything done, and it would be intrusive if our conversations with other people had to be punctuated by our conversations with our machines. Most seriously, even if speech recognition systems worked perfectly (and they don't), the result is no better than if the commands had been typed. So much of the frustration in using a computer is not the effort to enter the commands, it's figuring out how to tell it to do what you want, or trying to interpret just what it has done. Speech is a piece of the puzzle, but it doesn't address the fundamental mysteries confronting most computer users.

A dream interface has always been dreams, using mind control to direct a computer. There is now serious work being done on making machines that can read minds. One technique used is magnetoencephalography (MEG), which places sensitive detectors of magnetic fields around a head and measures the tiny neural currents flowing in the brain. Another technique, functional magnetic resonance imaging, uses MRI to make a 3D map of chemical distributions in the brain to locate where metabolic activity is happening. Both of these can, under ideal conditions, deduce something about what is being thought, such as distinguishing between listening to music and looking at art, or moving one hand versus the other. The problem that both struggle with is that the brain's internal representation is not designed for external consumption.

Early programmers did a crude form of MEG by placing a radio near a computer; the pattern of static could reveal when a program got stuck in a loop. But as soon as video displays came along it became much easier for the computer to present the information in a meaningful form, showing just what it was doing. In theory the same information could be deduced by measuring all of the voltages on all of the leads of the chips; in practice this is done only by hardware manufacturers in testing new systems, and it takes weeks of effort.

Similarly, things that are hard to measure inside a person are simple to recognize on the outside. For example, hold your finger up and wiggle it back and forth. You've just performed a brain control task that the Air Force has spent a great deal of time and money trying to replicate. They've built a cockpit that lets a pilot control the roll angle by thinking; trained operators on a good day can slowly tilt it from side to side. They're a long way from flying a plane that way.

In fact, a great deal of the work in developing thought interfaces is actually closer to wiggling your finger. It's much easier to accidentally measure artifacts that come from muscle tension in your forehead or scalp than it is to record signals from deep in the brain. Instead of trying to teach people to do the equivalent of wiggling their ears, it's easier to use the parts of our bodies that already come wired for us to interact with the world.

Another leading contender for the next big interface is 3D graphics. Our world is three-dimensional; why limit the screen to two dimensions? With advances in the speed of graphical processors it is becoming possible to render 3D scenes as quickly as 2D windows are now drawn. A 3D desktop could present the files in a computer as the drawers of a file cabinet or as a shelf of books, making browsing more intuitive. If you're willing to wear special glasses, the 3D illusion can be quite convincing.

A 3D display can even be more than an illusion. My colleague Steve Benton invented the reflection holograms on your credit cards; his group is now developing real-time holographic video. A computer calculates the light that would be reflected from a three-dimensional object, and modulates a laser beam to produce exactly that. Instead of tricking the eyes by using separate displays to produce an illusion of depth, his display actually creates the exact light pattern that the synthetic object would reflect.

Steve's system is a technological tour de force, the realization of a long-standing dream in the display community. It's also slightly disappointing to many people who see it, because a holographic car doesn't look as good as a real car. The problem is that reality is just too good. The eye has the equivalent of many thousands of lines of resolution, and a refresh rate of milliseconds. In the physical world there's no delay between moving an object and seeing a new perspective. Steve may someday be able to match those specifications with holographic video, but it's a daunting challenge.

Instead of struggling to create a computer world that can replace our physical world, there's an alternative: augment it. Embrace the means of interaction that we've spent eons perfecting as a species, and enhance them with digital content.

Consider Doug Engelbart's influential mouse. It is a two-dimensional controller that can be moved left and right, forward and backward, and intent is signaled by pressing it. It was preceded by a few centuries by another two-dimensional controller, a violin bow. That, too, is moved left and right, forward and backward, and intent is communicated by pressing it. In this sense the bow and mouse are very similar. On the other hand, while a good mouse might cost $10, a good bow can cost $10,000. It takes a few moments to learn to use a mouse, and a lifetime to learn to use a bow. Why would anyone prefer the bow?

Because it lets them do so much more. Consider the differences between the bow technique and the mouse technique:

Bow Technique	Mouse Technique
Sul ponticello (bowing close to the bridge)	click
Spiccato (dropping the bow)	double click
Martelé (forcefully releasing the stroke)	drag
Jeté (bouncing the bow)	
Tremolo (moving back and forth repeatedly)	
Sul tasto (bowing over the fingerboard)	
Arpeggio (bouncing on broken chords)	
Col legno (striking with the stick)	
Viotti (unaccented then accented note)	
Staccato (many martelé notes in one stroke)	
Staccato volante (slight spring during rapid staccato)	
Détaché (vigorous articulated stroke)	
Legato (smooth stroke up or down)	
Sautillé (rapid strike in middle of bow)	
Louré (separated slurred notes)	
Ondulé (tremolo between two strings)	

There's much more to the bow than a casual marketing list of features might convey. Its exquisite physical construction lets the player perform a much richer control task, relying on the intimate connection between the dynamics of the bow and the tactile interface to the hand manipulating and sensing its motion. Compare that nuance to a mouse, which can be used perfectly well while wearing mittens.

When we did the cello project I didn't want to ask Yo-Yo to give up this marvelous interface; I retained the bow and instead asked the computer to respond to it. Afterward, we found that the sensor

I developed to track the bow could respond to a hand without the bow. This kind of artifact is apparent any time a radio makes static when you walk by it, and was used back in the 1930s by the Russian inventor Lev Termen in his Theremin, the musical staple of science-fiction movies that makes eerie sounds in response to a player waving their arms in front of it.

My student Josh Smith and I found that lurking behind this behavior was a beautiful mathematical problem: given the charges measured on two-dimensional electrodes, what is the three-dimensional distribution of material that produced it? As we made headway with the problem we found that we could make what looks like an ordinary table, but that has electrodes in it that create a weak electric field that can find the location of a hand above it. It's completely unobtrusive, and responds to the smallest motions a person can make (millimeters) as quickly as they can make them (milliseconds). Now we don't need to clutter the desk with a rodent; the interface can disappear into the furniture. There's no need to look for a mouse since you always know where to find your hand.

The circuit board that we developed to make these measurements ended up being call a "Fish," because fish swim in 3D instead of mice crawling in 2D, and some fish that live in murky waters use electric fields to detect objects in their vicinity just as we were rediscovering how to do it. In retrospect, it's surprising that it has taken so long for such an exquisite biological sense to get used for computer interfaces. There's been an anthropomorphic tendency to assume that a computer's senses should match our own.

We had trouble keeping the Fish boards on hand because they would be carried off around the Media Lab by students who wanted to build physical interfaces. More recently, the students have been acquiring as many radio-frequency identification (RFID) chips as they can get their hands on. These are tiny processors, small enough even to be swallowed, that are powered by an external field that can also exchange data with them. They're currently used in niche applications, such as tracking laboratory animals, or in the key-chain tags that enable us to pump gas without using a credit card. The students use them everywhere else. They make coffee cups that can tell the coffeemaker how you like your coffee, shoes that can tell a doorknob who you are, and mouse pads that can read a Web URL from an object placed on it.

You can think of this as a kind of digital shadow. Right now objects live either in the physical world or as icons on a computer screen. User interface designers still debate whether icons that appear to be three-dimensional are better than ones that look two-dimensional. Instead, the icons can really become three-dimensional; physical objects can have logical behavior associated with them. A business card should contain an address, but also summon a Web page if placed near a Web browser. A pen should write in normal ink, but also remember what it writes so that the information can be recalled later in a computer, and it should serve as a stylus to control that computer. A house key can also serve as a cryptographic key. Each of these things has a useful physical function as well as a digital one.

My colleague Hiroshi Ishii has a group of industrial designers, graphical designers, and user interface designers studying how to build such new kinds of environmental interfaces. A recurring theme is that interaction should happen in the context that you, rather than the computer, find meaningful. They use video projectors so that tables and floors and walls can show relevant information; since Hiroshi is such a good Ping-Pong player, one of the first examples was a Ping-Pong table that displayed the ball's trajectory in a fast-moving game by connecting sensors in the table to a video projector aimed down at the table. His student John Underkoffler notes that a lamp is a one-bit display that can be either on or off;

John is replacing lightbulbs with combinations of computer video projectors and cameras so that the light can illuminate ideas as well as spaces.

Many of the most interesting displays they use are barely perceptible, such as a room for managing their computer network that maps the traffic into ambient sounds and visual cues. A soothing breeze indicates that all is well; the sights and sounds of a thunderstorm is a sign of an impending disaster that needs immediate attention. This information about their computer network is always available, but never demands direct attention unless there is a problem.

Taken together, ambient displays, tagged objects, and remote sensing of people have a simple interpretation: the computer as a distinguishable object disappears. Instead of a fixed display, keyboard, and mouse, the things around us become the means we use to interact with electronic information as well as the physical world. Today's battles between competing computer operating systems and hardware platforms will literally vanish into the woodwork as the diversity of the physical world makes control of the desktop less relevant.

This is really no more than Piaget's original premise of learning through manipulation, filtered through Papert and Kay. We've gotten stuck at the developmental stage of early infants who use one hand to point at things in their world, a decidedly small subset of human experience. Things we do well rely on all of our senses.

Children, of course, understand this. The first lesson that any technologist bringing computers into a classroom gets taught by the kids is that they don't want to sit still in front of a tube. They want to play, in groups and alone, wherever their fancy takes them. The computer has to tag along if it is to participate. This is why Mitch Resnick, who has carried on Seymour's tradition at the Media Lab, has worked so hard to squeeze a computer into a Lego brick. These bring the malleability of computing to the interactivity of a Lego set.

Just as Alan's computer for kids was quickly taken over by the grown-ups, Lego has been finding that adults are as interested as kids in their smart bricks. There's no end to the creativity that's found expression through them; my favorite is a descendent of the old LOGO turtle, a copier made from a Lego car that drives over a page with a light sensor and then trails a pen to draw a copy of the page.

A window is actually an apt metaphor for how we use computers now. It is a barrier between what is inside and what is outside. While that can be useful at times (such as keeping bugs where they belong), it's confining to stay behind it. Windows also open to let fresh air in and let people out.

All along the coming interface paradigm has been apparent. The mistake was to assume that a computer interface happens between a person sitting at a desk and a computer sitting on the desk. We didn't just miss the forest for the trees, we missed the earth and the sky and everything else. The world is the next interface.

HOW

. . . will things that think be developed?

by taking advantage of what nature
already knows how to do

✦

through research that maximizes
contact, not constraints

✦

with education that happens as it is needed,
rather than in advance

✦

by interconnecting systems of people and
things to solve hard problems

The Nature of Computation

Information technology can be thought of as a pyramid. At the top are the supercomputers, the biggest, fastest machines, bought by the biggest, richest institutions. Then come the large mainframes that run companies. After that are the servers for a department, high-powered workstations for demanding applications, and then the common PC. As the cost goes down, the number of these things in use goes up.

Below the PC are emerging devices that promise to be used in still larger numbers. A few hundred dollars buys a Personal Digital

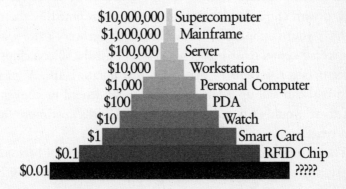

Assistant, a pocket-sized computer handy for mobile information access. For tens of dollars, the wristwatch has for many years been the one processor and display that people carry around with them. It is now acquiring new capabilities, such as a pager, or a GPS receiver, or the Swatch Access watches used as electronic ski-lift tickets. Europe is also leading the way with smart cards, computers the size of a credit card that are used much the same way, but that store electronic cash instead of requiring access to a remote computer. They were first used to eliminate the need for change to use pay phones, and are now spreading to all areas of electronic commerce. Finally, for tens of cents come the RFID chips that are used to tag things with an electronic identity. These are used in the badges that control access to a workplace, and in tracking laboratory animals. And that's where it stops. No one's been able to fabricate, package, and sell chips for less than about ten cents. While that might not appear to be a severe limit, think about what it leaves off: just about everything!

One of the most frequent challenges presented to me is finding a way to compute for pennies. Companies have a need to track the identity, location, and environment of the things they make, but in competitive markets they can't afford to add more than a few cents per item. While it would be nice to be able to walk out of a supermarket without waiting in line because the groceries come tagged with chips that can automatically be scanned by the door, or have your cupboard at home contact the grocery store when it needs restocking, these things won't happen if the silicon chips add more than a few cents to the cost of the potato chips. A business card that can call up a business's Web page would be convenient, but there wouldn't be a business left if its cards cost more than a few cents.

This is such an intensely commercial question that it becomes an interesting scientific one. It's not possible to reach one-cent computers just by cutting corners in the fabrication of conventional chips; the whole industry has been trying unsuccessfully to do that for many years. The only way to get there is to be smarter about taking advantage of what nature already knows how to do.

A clue is provided by the cheapest digital device of all, the shoplifting tag. This is a one-cent one-bit remotely read memory device, storing the status of a product (purchased vs. stolen). Many of these, such as the white rectangular strip that you find on a compact disk, are made by a company called Sensormatic. These contain a metallic glass, a thin sheet of a metal alloy that was cooled so quickly when it formed that the atoms were frozen in a random arrangement. It looks like a piece of aluminum foil, but when it is exposed to an oscillating magnetic field it flexes slightly and then rings much like a tuning fork. This resulting oscillation is much too fast to be audible, but it in turn creates an oscillating magnetic field that can be detected. The portals around the door at a CD store contain the coils for exciting and detecting these oscillations.

My student Rich Fletcher was working with Sensormatic on squeezing more functionality into these materials by trimming them to varying lengths so that they would oscillate at different frequencies, and sensing the motion of magnets near them by how they shift the frequency. This let him store a few bits in the tag, and use it as a passive wireless user-interface controller. By putting the coils in a table instead of a door, Rich could create a work space that could distinguish among objects brought in, and detect how they were moved. Doing that isn't remarkable, but doing it for a penny is. At that cost this capability could be put into most anything, suggesting that smart materials might be able to replace dumb chips. Indeed, Rich has since found many other materials that can be remotely sensed to determine their identity, track their position, and detect properties such as pressure and temperature.

This approach rapidly runs into a limit on the amount of information that can be stored in a tag because there are not enough distinguishable frequencies. Very little data is contained in the call sign of a radio station; the interesting part is the signal the station broadcasts. For a smart material to be able to send out a more complex signal it needs to be nonlinear. If you hit a tuning fork twice as hard it will ring twice as loud but still at the same frequency. That's a linear response. If you hit a person twice as hard they're unlikely just to shout twice as loud. That property lets you learn more about the person than the tuning fork.

I set out to find a nonlinearity that we could use to store more data in smart material tags. To keep the cost down to a penny it had to be a natural property of the material rather than something that would require elaborate fabrication. An intriguing candidate is the interaction between the atoms in a material. Chemists study these with the technique of nuclear magnetic resonance (NMR). Magnetic resonance imaging (MRI) uses exactly the same mechanism, minus the word "nuclear," which might scare patients.

NMR is performed in much the same way as shoplifting tags are detected, applying an oscillating magnetic field and then listening for the resulting oscillating field. In the tag the oscillation came from a mechanical vibration of the material; in NMR the oscillation arises from the nuclei of the atoms. In a magnetic field these precess like tiny spinning tops. The rate at which they precess depends on the orientation of neighboring nuclei, spinning faster if they point in the same direction for example. By probing these neighboring interactions with NMR, chemists are able to deduce the structure of molecules. MRI applies a magnetic field that varies over space, so that the oscillation frequency of a nucleus becomes a function of its position.

The more I learned about NMR, the more puzzled I became. It usually is performed by putting a tube full of a liquid in a strong magnetic field. A coil wrapped around the tube sends in an oscillating magnetic field to make the nuclei precess, and then detects the resulting magnetic field from the response of the nuclei. The chemists analyze NMR by using the language of quantum mechanics. Quantum mechanics describes the bizarre behavior of very small things, not a macroscopic test tube. Quantum mechanically there is nothing preventing this book from splitting into ten copies and leaping to the moon, but the interactions with you and the rest of the world force it to remain sensibly classical. To understand the paradox of how an apparently classical test tube can behave quantum mechanically, I started lugging around a distinctly macroscopic NMR bible that chemists use, a book by Ernst.

I was not getting anywhere until I was forced to do jury duty. For months I had been trying to avoid serving, but I ran out of excuses and had to clear a few days and report to the Cambridge courthouse. There I was locked in the jury room for a day with nothing to do but read Ernst. Freed from all my usual distractions, I finally was able to understand just how elegant the apparently impenetrable and self-referential chemists' language is.

In NMR each molecule in the liquid behaves as an independent but identical quantum system. The experiments treat the liquid as a single effective molecule that in reality is represented by an enormous number of redundant copies. All of the classical interactions needed to hold and interact with the liquid do ruin some of those copies, but there are so many of them that the quantum mechanical properties of the rest can persist for many seconds, plenty of time to perform an experiment. This means that NMR provides a convenient way to access a nonlinear quantum mechanical system.

Those attributes happen to be exactly what's needed for the future of high-performance computing. The search for cheap chips was leading to something much more. We've come to expect that every year or so our computers will double in speed and capacity.

This trend was first noticed by Gordon Moore of Intel in 1965, and has continued for the last three decades. It's not a law of nature; it's an observation about remarkable engineering progress. From the very beginning it was clear where this was heading. Each year it appears that it will be hard to continue improving performance, and each year some new tricks are found to squeeze out the next round. But some time around 2020 or so everything will hit bottom. At the rate of current progress, the wires will be one atom wide, the memory cells will have one electron, and the fab plant will cost the GNP of the planet so that no one can afford to build it anyway. Further improvements cannot come from the current path of shrinking silicon circuits.

In part, this doesn't matter because most of the current constraints on using a computer have nothing to do with the speed of the processor, they come from the user interface, or the power requirement, or the software. A few of the billion dollars spent on building chip fabs could profitably be used elsewhere. But in part it does matter, because there are some hard problems that always demand faster computers.

Finding prime factors, for example, is the mathematical problem that underpins the cryptography used for electronic commerce. The difficulty of factoring is what lets you send a credit card number from a Web browser without worrying about an eavesdropper obtaining it. A desktop computer can multiply together the factors of a four-hundred-digit number in an instant, but finding them would take today's fastest computer the age of the universe. That is because factoring is an exponential algorithm, meaning that doubling the size of the number to be factored squares instead of doubles the time needed to do it. And beyond factoring, it's also true that builders of each generation of computers have confidently predicted that they were sufficient for most applications, only to find entirely new uses for new machines that weren't anticipated by extrapolating the old ones. I don't feel limited because my desktop computer isn't one thousand times faster, but that reflects a poverty of my imagination as much as an observation about what I think I need a computer for.

There aren't many places to turn for alternatives to speed up computing. Light travels a foot in a billionth of a second, the cycle time of today's fastest chips. This means that they cannot be clocked much faster because there isn't time for the signals to go anywhere.

Another option is to keep the clock speed but use more computers in parallel. Not only would this require an annual doubling of the number of computers to keep pace with Moore's law, parallelism alone does not scale very well. Scientists have managed to perform logical functions with DNA molecules, turning a tube full of DNA into an enormous number of computers, about 10^{23} of them. That's a lot. Because they communicate with each other slowly, the easiest way to use them is like a lottery, preparing all possible answers to a problem and then checking to see which molecule has the winning solution. But since the number of possible factors of a cryptographic key is exponential in the size of the key, this means that 10^{23} computers could simultaneously check for the factors of just a twenty-three-digit number.

There's only one thing in our universe that is exponentially big, but is not used for computing: quantum mechanics. A classical bit can be either a 0 or a 1. A bit governed by the laws of quantum mechanics can simultaneously be a 0 or a 1. To describe it you need to specify the probability that you'll find a 0 if you measure it, and the probability that you'll find a 1. The state of a classical computer with N bits is specified by giving those bits, but for a quantum computer it requires giving 2^N coefficients, one for each possible combination of the bits. For just fifty quantum bits, $2^{50} = 10^{15}$, storing more information than the largest classical computers. Because a quantum computer could perform all possible calculations at the same time, it might be the key to building more powerful computers.

This question was first asked in the early 1980s by Richard Feynmann at Caltech, Paul Benioff at Argonne National Laboratory, and David Deutsch at Oxford University. Feynmann was struck by the recognition that it takes a classical computer an exponential amount of time to model a quantum system, because it has to keep track of all of the possible states of the system in parallel. He wondered if a computer governed by the laws of quantum mechanics might be able to efficiently model other quantum systems. Feynmann's conjecture, recently supported by Seth Lloyd at MIT, had broader implications than he realized.

The study of quantum computing remained a theoretical curiosity through the 1980s, pursued by an unusual collection of people who had to know a great deal about both physics and computer science, and also be slightly strange. Meetings of this community were a delight because the people involved were so unusual, and the discussion so irrelevant to any practical concerns. That began to change in the early 1990s. A series of results were proved by David Deutsch, Richard Josza of Plymouth University, and Dan Simon, now at Microsoft, showing that a quantum computer is more powerful than a classical computer for a series of increasingly less trivial problems. In 1994 Peter Shor of AT&T was able to use these techniques to show that a quantum computer could find prime factors in polynomial instead of exponential time. This made factoring a four-hundred-digit number almost as easy as multiplying its factors.

These results drew on a second aspect of quantum mechanics that is even more bizarre than the ability to be in many states simultaneously. If a quantum bit is in a superposition of 0 and 1, and it interacts with a second bit, the value of the second bit will

depend on the state of the first. If they're now separated to opposite ends of the universe and the first bit is measured, it is forced
to decide between being a 0 or a 1. At that instant the value of the
second bit is determined. This appears to be an instantaneous action at a distance, something that made Einstein very unhappy. It
is called entanglement, and it serves in effect to wire together the
bits in a quantum computer. Without entanglement a quantum
computer would have the same problem as a DNA computer, trying many answers at the same time and then having to locate the
correct one, like trying to find a needle in a haystack. With entanglement, a single quantum computer can be certain to solve a factoring problem.

Naturally, the three-letter agencies (NSA, CIA, . . .) panicked.
Here was a very real threat to information security, coming from
an entirely unexpected quarter. Since by that time the result was
already widely known they couldn't cover it up, but they could try
to keep ahead of the competition. So they started showing up at
meetings, in effect offering a purchase order to anyone who would
build them a quantum computer. They started to relax when no
one serious would take one. In the next year many people realized
just how hard it was going to be to build a practical quantum
computer. It would need to be carefully isolated from the rest of
the world so that it could stay quantum, but would need to be accessible to the outside world so that the initial conditions could be
loaded in, the computation run forward, and the answer read out.
Papers started to appear saying that quantum computers could not
be built.

Then, in 1996, two things happened to scare the three-letter
agencies again. The first was the discovery by Peter Shor, Charles
Bennett of IBM, and Andy Steane of Oxford, that quantum error
correction is possible. Just as classical computers add extra bits to
memories to catch mistakes, they realized that quantum computers

can use extra quantum bits to fix their errors. This means that an imperfect quantum computer could perform a perfect computation, as long as it is just good enough to be able to implement the error correction.

Here's where NMR comes in. After my day of jury duty I understood that it might be possible to solve the paradox of connecting the classical and quantum worlds with the chemist's trick. Instead of working with a single computer, if a tube full of quantum computers in the form of a liquid sample in an NMR apparatus was used, then all of the external interactions might ruin a few of them, but since each effective quantum bit would be represented in an enormous number of molecules it would retain its quantum coherence for a much longer time. For this idea to work it would have to use natural molecules, since it's not feasible to individually fabricate that many computers. So the question then became whether a molecule could compute.

Classically, a digital computer must be able to flip a bit (the NOT gate), and perform a logical operation on two bits (such as the AND gate). Any other logical operation can be assembled from these components. Similarly, a quantum computer must be able to rotate a quantum bit to any angle between 0 and 1, and must be able to perform a nonlinear operation between two bits. These are easy to do with NMR. The precession of a nucleus in a magnetic field is exactly the kind of rotation that is needed, and the interaction between the nuclei provides the nonlinearity.

I arrived for a visit at the Institute for Theoretical Physics in Santa Barbara, armed with this knowledge of how to represent quantum information in what appears to be a classical system, and how to apply magnetic fields to manipulate the quantum information. On the first day I stopped by to see Isaac Chuang (now at IBM), and emerged from his office a week later. He had just finished working out how to reduce a quantum computer to a series of equivalent operations. The operations that he needed and that I knew how to do matched up perfectly.

By the end of the week we had worked out the details of programming a molecular quantum computer. We had an eerie feeling that we weren't really discovering anything, just lifting a curtain to reveal a beautiful structure that had been there all along waiting for someone to notice it.

A liquid is in fact that most perfect system that could be built for quantum computing. It's easy to obtain in huge quantities, the molecules all come perfectly assembled without any manufacturing imperfections, and the cloud of electrons around the nucleus and the rapid tumbling of the molecules in the liquid protect the quantum coherence of the nuclei. Once we understood the connection between liquid NMR and quantum computing we didn't even need to do a demonstration experiment, because the chemists had been doing them for years. Their molecular studies were showing all of the working elements of a quantum computer, without their realizing that's what they were doing.

Richard Feynmann gave a seminal talk in 1959 entitled "There's Plenty of Room at the Bottom." He foresaw the problems coming in shrinking computers, and wondered if we could jump to making molecular computers. This inspired the field of nanotechnology that sought to do just that. Unfortunately, the study of nanotechnology has produced thoughtful analyses and compelling visions, but very little in the way of working machines. It now turns out that ordinary molecules all along have known how to compute—we just weren't asking them the right questions.

This was one of those ideas that was ready to be thought. An MIT/Harvard group found a related way to manipulate quantum information in a liquid, one at Los Alamos found still another, and many groups around the world are now using these techniques to make small quantum computers. The first experimental computation we did that required fewer steps than on a classical computer was a search problem, using the molecule chloroform. Given no advance knowledge, unlocking a classical padlock with N settings requires about $N/2$ attempts. Lov Grover at Bell Labs has shown that this can be done in $\sqrt{N}$ steps on a quantum computer. Our chloroform computer had four states; our quantum search was able to find in a single step the answer that took a few tries classically.

Daunting challenges remain if these demonstration experiments are ever to grow up to exceed the performance of today's fastest classical computers, although these are increasingly technological rather than conceptual problems. The most severe limit is a rapid decrease in sensitivity for larger molecules, but this might be solved by using lasers to align the nuclei. The quantum coherence lasts for thousands of operations, approaching the limit needed to be able to use quantum error correction. This also needs a technique such as the optical nuclear alignment, to prepare the extra quantum bits used in the error correction. Finally, as the size of a molecule is increased, the interactions between nuclei at the far ends of a molecule become too weak to use for computing, but Seth Lloyd has shown that a polymer with a simple repeating atomic pattern is almost as useful and requires that the nuclei interact only with their nearest neighbors.

These scaling constraints offer a glimpse of an ideal molecule for quantum computing. It should have a regular structure, so that Seth's technique can be used. Better still, the pattern should be 2D or 3D rather than a linear chain, to reduce the time to pass messages around the molecule. It should have a rigid configuration, so that all of the local atomic environments are identical, and it should have a special structure at the end that can be used to load in data and read out results. This sounds disturbingly like a microtubule.

Microtubules are a kind of molecular scaffolding in neurons.

While there's no compelling reason to believe that quantum mechanics has anything to do with consciousness, those who do think that it does have come to identify microtubules as the seat of quantum consciousness. I couldn't begin to guess how the brain could perform the kinds of molecular manipulations that we do in NMR, although it is amusing to observe that a head in an MRI magnet is a perfectly good medium in which to perform a quantum computation.

The brain may not be the best place to look for guidance in how to build a quantum computer, but it does have a valuable lesson to teach us about how to build more powerful computers. Moore's law demands exponential improvements in performance. Quantum mechanics promises one way to do that; analog circuitry offers another.

The neurons in a brain are about a million times slower than the transistors in a computer, yet they can instantly perform tasks that tax the fastest computers. There are many more neurons in the brain than transistors in a computer, about 10^{11} of them, and even more connections among them, roughly 10^{15} synapses. These synapses are the key.

Many computing problems reduce to finding good ways to approximate complicated functions. For example, for a computer to recognize faces it must have a function that can take as an input the pixels in an image and provide as an output the name associated with the face. There are many techniques for doing this, but they all reduce in the end to some kind of function that relates inputs to outputs. In traditional engineering practice unknown functions are represented by using some family of basis functions, such as a collection of polynomial powers or trigonometric functions, combined with a set of weighting coefficients. There are relatively straightforward techniques for finding the best coefficients to match a given set of data.

This isn't how the brain does it. In the brain the coefficients correspond to the synaptic connection strengths, and the basis functions are the S-shaped response of the neurons. The brain uses the coefficients in forming the arguments of the basis functions, rather than using them just to combine the results. Engineers have historically stayed away from doing this because it can be shown that there's no longer a practical procedure to find the best values for the coefficients. All that can be done is some kind of search that learns reasonable ones.

The study of neural networks was inspired by the goal of emulating the organization of the brain. While the brain is still far too complex to model in any detail, simple mathematical networks that put the unknown coefficients before the nonlinear basis functions were surprisingly successful. It is now understood why.

A characteristic feature of difficult recognition problems is that there may be a huge number of inputs, such as all of the pixels in an image. In a traditional model, one coefficient is needed for each possible combination of inputs and basis functions. This rapidly explodes, so that if ten are needed for one input, one hundred are needed for two, one thousand for three, and so forth. This is called the "curse of dimensionality." A neural network model is much more flexible. The adjustable coefficients inside the nonlinear functions are far more powerful; the number required is a polynomial function instead of an exponential function of the number of inputs.

Even so, there's still the problem of finding their values. A second result comes to the rescue here. It turns out that these models are so flexible, almost anywhere they start out there is a good solution nearby. A search procedure does not need to find the single best solution, just an acceptable one, and there is a huge number of them. This is ensured by using a large number of neurons, far more than would be considered sane in a conventional model. This property has come to be called the "blessing of dimensionality." It helps explain why learning is so essential to how the brain functions.

So here is the second exponential insight: use models that require learning the values of coefficients that appear inside nonlinear functions. While this is not applicable to all computational problems, it works for just the sorts of things that people do well, such as recognizing patterns and prioritizing tasks. A digital computer can simulate this kind of architecture, but it's a misplaced application of the digital precision. Simpler analog circuits can do the same thing with far fewer parts.

Analog circuits were used in the early days of computing, but they fell out of favor because they accumulate errors. Digital computers can correct their mistakes and hence perform arbitrarily long computations. However, this is not an essential distinction between analog and digital systems. Ultimately everything is analog; the apparently digital values are a consequence of carefully designed analog circuits. We're now learning how to marry the most desirable features of both worlds, correcting analog errors while using the analog flexibility to squeeze much more performance out of a simple physical system.

For example, Geoff Grinstein of IBM and I studied what are called spread-spectrum coding techniques, which are used to intentionally make a signal appear to be more random than it is. While this may appear to be a perverse aim, it turns out that almost every property of engineered systems is improved when they work with effectively random signals (more people can share a communications channel with less interference, more data can be stored on a disk, and so forth). There is a beautiful theory for how to generate bits for use in spread spectrum that appear to be completely random but in fact are entirely predictable. A transmitter uses such a sequence to spread its signal and the receiver then uses an identical copy to recover the message (this idea can be traced back to the actress Hedy Lamarr and composer George Antheil using piano rolls in World War II). This is what you hear when your modem hisses. The catch is that if the receiver is remote from the transmitter it has a difficult computing job to do to figure out the spreading sequence that the transmitter used. This is part of what a GPS receiver must do to lock onto a satellite and find its position.

Two copies of almost any analog system, if allowed to interact in almost any way, have a very interesting property: they synchronize. This used to be seen in the simultaneous ticking of mechanical clocks on the wall of a clock shop. Geoff and I wondered if we could use this property to solve the problem of spread-spectrum acquisition. Much to our pleasant surprise, we discovered that we could design simple continuous systems that exactly matched the response of the familiar digital ones for digital signals, but if they started in the wrong state they could use their analog freedom to synchronize onto the digital signal. Instead of splitting the job into an analog detector and then a digital search program as is done now, a physical analog part can do everything.

The digital world has gone on refining Turing's machine, when even Turing understood its limits. One more area that he made a seminal contribution to was explaining how biology can form patterns by using chemical reactions. This useful behavior is easy to find in nature but hard to model on a digital computer. Now that we're approaching the end of the era of rapid increases in the performance of digital computers, like Turing we have no choice but to pay more attention to information processing in the analog world.

When Feynmann said that there's plenty of room at the bottom, he meant that there was a long way to go to shrink computers down to atomic sizes. That's no longer true. With difficulty we are approaching that limit with carefully engineered circuits, and can now glimpse how to eliminate all that effort by using natural materials. Although he wasn't thinking about economics, that may carry the most interesting implication of all of his challenge. There's still plenty of room at the bottom to make much cheaper and more widely accessible computers. Nature is not only good at designing computers, it has a few lessons left to teach us about manufacturing them.

The Business of Discovery

There used to be quite a demand for psychic mediums who would go into "spirit cabinets" and channel fields to contact lost souls who would communicate by making sounds. In my lab, once we had developed the techniques to induce and measure weak electric fields around the human body, we realized that we could make a modern-day spirit cabinet (well, everything but the lost souls part). So of course we had to do a magic trick, which we were soon working on with the magicians Penn & Teller, who were planning a small opera with Tod Machover. We expected to have fun; we didn't expect to save the lives of infants, or learn something about how industry and academia can remove technological barriers by removing organizational barriers.

The hardware for our spirit cabinet was developed by Joe Paradiso, a former experimental particle physicist who in his spare time builds electronic music synthesizers that look just like enormous particle detectors. Joe designed a seat that radiates a field out through the medium's body, and from that can invisibly measure the smallest gestures. Tod shaped a composition for it, and Penn & Teller toured with it.

The trick worked technically, but it ran into an unexpected artistic problem: audiences had a hard time believing that the performers created rather than responded to the sounds. Science fiction had prepared them to believe in hyperspace travel but not in a working spirit cabinet. It was only by sitting people in the chair that we could convince them that it did work; at that point we had a hard time getting them back out of it.

I saw this project as an entertaining exercise for the physical models and instrumentation that we were developing to look at fields around bodies, not as anything particularly useful. The last thing I expected was to see it turn into an automotive safety product. But after we showed it, I found a Media Lab sponsor, Phil Rittmueller from NEC, in my lab telling me about child seats. NEC makes the controllers for airbags, and they were all too familiar with a story that was about to hit the popular press: infants in rear-facing child seats were being injured and killed by airbags. Deaths were going to result from leaving the airbags on, and also from switching them off. Finding a way for a smart airbag to recognize and respond appropriately to the seat occupant was a life-and-death question for the industry.

No one knew how to do it. To decide when to fire, the seat needed to distinguish among a child facing forward, a child facing reverse, a small adult, and a bag of groceries. Under pressure to act, the National Highway Traffic Safety Administration was about to mandate a standard for disabling the airbag based on the weight of the occupant, an arbitrary threshold that would be sure to make mistakes. The only working alternative imposed the unrealistic requirement that all child seats had to have special sensor tags installed in them.

Phil wondered if our fancy seat could recognize infants as well as magicians. My student Josh Smith veered off from what he was doing for a few weeks to put together a prototype, and to our very pleasant surprise it not only worked, it appeared to outperform anything being considered by the auto industry. After NEC showed the prototype to some car companies they wanted to know when they could buy it. NEC has since announced the product, the Passenger Sensing System.

It looks like an ordinary automobile seat. Flexible electrodes in the seat emit and detect very weak electric fields, much like the weak fields given off from the cord of a stereo headphone. A controller interprets these signals to effectively see the three-dimensional configuration of the seat occupant, and uses these data to determine how to fire the airbag. The beauty of the system is that it's invisible and requires no attention; all the passenger has to do is sit in the seat. Although the immediate application is to disable the airbag for rear-facing infant seats, in the future the system will be used to control the inflation of an airbag based on the location of the occupant, and more generally help the car respond to the state of the occupant.

Now I would never have taken funding for automobile seat safety development—it's too remote from what I do, and too narrow. NEC would never have supported magic tricks internally—it's too remote from what they do, and too crazy. Yet by putting these pieces together, the result was something unambiguously useful that we probably could not have gotten to any other way. This is one of the secrets of how the Media Lab works with industrial sponsors: maximize contact, not constraints.

That's what much of academia and industry carefully prevent. The organization of research and development in the United States can be directly traced to an influential report that Vannevar Bush wrote for Franklin Roosevelt in 1945.

Two technologies developed during World War II arguably ended the conflict, first radar and then nuclear bombs. These were created under the auspices of the then-secret Office of Scientific Research and Development, directed by Vannevar Bush. After the war President Roosevelt asked him to figure out how to sustain that pace of development for peacetime goals, including combating disease and creating jobs in new industries.

The resulting report, *Science—The Endless Frontier,* argued that the key was government support of basic research. Both nuclear weapons and radar were largely a consequence of fundamental studies of the subatomic structure of matter, the former directly from nuclear physics, and the latter through the instruments that had been developed for nuclear magnetic resonance experiments. Just when science and technology emerged as a key to economic competitiveness after the war, the country was facing a shortage of scientists and engineers who could do the work. Attracting and training them would be essential.

Vannevar Bush proposed that the government's postwar role should be to fund curiosity-driven basic research studies. These should be done externally, at universities and research institutes, and be evaluated solely with regard to their scientific merit without any concern for practical applications. Since the direction of true basic research can never be predicted, the government would provide long-term grants, awarded based on the promise and record of a research project rather than claims of expected outcomes of the work. Researchers doing solid work could expect steady funding.

The useful fruits of the basic research would then be picked up by applied research laboratories. Here the government would have a stronger role, both through the work of its own agencies, and through fostering a regulatory climate that encouraged industry to do the same. Finally, the applied results would move to industry where product development would be done.

He proposed the creation of a new agency to support all government research, the National Research Foundation. By the time the enabling legislations was passed in 1950 the title had changed to the National Science Foundation (NSF) since there were too many entrenched government interests unwilling to cede control in areas other than basic science. Vannevar Bush thought that a staff of about fifty people and a budget of $20 million a year should be sufficient to do the job.

In 1997 the NSF had twelve hundred employees and a budget of $3 billion a year. Attracting new scientists is no longer a problem; finding research jobs for them is. The NSF gets so many proposals from so many people that simply doing great work is no longer sufficient to ensure adequate, stable, long-term funding. Vannevar Bush's system is straining under the weight of its own successful creation of an enormous academic research establishment.

It's also struggling to cope with the consequences of the growth of the rest of the government. The original proposal for the National Research Foundation recognized that research is too unlike any other kind of government function to be managed in the same way; grantees would need the freedom to organize and administer themselves as they saw fit. The usual government bidding and accounting rules would have to be relaxed to encourage basic research.

Not that the government has been particularly successful at overseeing itself. The shocking inefficiency of the federal bureaucracy led to the passage in 1993 of the Government Performance and Results Act. GPRA sought to bring industrial standards of accountability to bear on the government, forcing agencies to develop clear strategic plans with measurable performance targets, and basing future budgetary obligations on the value delivered to the agencies' customers.

This means that the U.S. government is now in the business of producing five-year plans with production targets, just as the former Soviet Union has given up on the idea as being hopelessly unrealistic. GPRA applies to the NSF, and its grantees, and so they must do the same. Therefore a basic research proposal now has to include annual milestones and measurable performance targets for the research. My industrial sponsors would never ask me to do that. If I could tell them exactly what I was going to deliver each year for the next five years then I wouldn't be doing research, I would be doing development that they could better perform internally.

Vannevar Bush set out to protect research from the pressure to deliver practical applications, freeing it to go wherever the quest for knowledge leads. GPRA seeks to nail research to applications, asking to be told up front what a project is useful for. Both are good-faith attempts to derive practical benefits from research funding, and both miss the way that so much innovation really happens.

Vannevar Bush's legacy was felt at Bell Labs when I was there before the breakup of the phone system in the 1980s. The building was split into two wings, with a neutral zone in the middle. Building 1 was where the basic research happened. This was where the best and the brightest went, the past and future Nobel laureates. The relevant indicator of performance was the number of articles in the prestigious physics journal that publishes short letters, a scientific kind of sound bite. Building 2 did development, an activity that was seen as less pure from over in Building 1. This didn't lead to Nobel prizes or research letters. Conversely Building 1, viewed from Building 2, was seen as arrogant and out of touch with the world. The cafeteria was between the buildings, but the two camps generally stuck to their sides of the lunchroom. Meanwhile, the business units of the telephone company saw much of the whole enterprise as being remote from their concerns, and set up their own development organizations. Very few ideas ever made it from Building 1 to Building 2 to a product.

There were still plenty of successes to point to, the most famous being the invention of the transitor. But this happened in exactly the reverse direction. The impetus behind the discovery of the transistor was not curiosity-driven basic research, it was the need to replace unreliable vacuum tubes in amplifiers in the phone system. It was only after the transistor had been found that explaining how it worked presented a range of fundamental research questions that kept Building 1 employed for the next few decades.

Few scientific discoveries ever spring from free inquiry alone. Take the history of one of my favorite concepts, entropy. In the mid-1800s steam power was increasingly important as the engine of industrial progress, but there was very little understanding of the performance limits on a steam engine that could guide the development of more efficient ones. This practical problem led Rudolf Clausius in 1854 to introduce entropy, defined to be the change in heat flowing into or out of a system, divided by the temperature. He found that it always increases, with the increase approaching zero in a perfect engine. Here now was a useful test to see how much a particular engine might be improved. This study has grown into the modern subject of thermodynamics.

The utility of entropy prompted a quest to find a microscopic explanation for the macroscopic definition. This was provided by Ludwig Boltzmann in 1877 as the logarithm of the number of different configurations that a system can be in. By analyzing the possible ways to arrange the molecules in a gas, he was able to show that his definition matched Clausius's. This theory provided the needed connection between the performance of a heat engine and the properties of the gas, but it also contained a disturbing loophole that Maxwell soon noticed.

An intelligent and devious little creature (which he called a demon) could judiciously open and close a tiny door between two sides of a box, thereby separating the hot from the cold gas molecules, which could then run an engine. Repeating this procedure would provide a perpetual-energy machine, a desirable if impossible outcome. Many people spent many years unsuccessfully trying to exorcise Maxwell's demon. An important step came in 1929 when Leo Szilard reduced the problem to its essence with a single molecule that could be on either side of a partition. While he wasn't able to solve the demon paradox, this introduced the notion of a "bit" of information.

Szilard's one-bit analysis of Maxwell's demon provided the inspiration for Claude Shannon's theory of information in 1948. Just as the steam engine powered the Industrial Revolution, electronic communications was powering an information revolution. And just as finding the capacity of a steam engine was a matter of some industrial import, the growing demand for communications links required an understanding of how many messages could be sent through a wire. Thanks to Szilard, Shannon realized that entropy could measure the capacity of a telephone wire as well as an engine. He created a theory of information that could find the performance limit of a communications channel. His theory led to a remarkable conclusion: as long as the data rate is below this capacity, it is possible to communicate without any errors. This result more than any other is responsible for the perfect fidelity that we now take for granted in digital systems.

Maxwell's demon survived until the 1960s, when Rolf Landauer at IBM related information theory back to its roots in thermodynamics and showed that the demon ceases to be a perpetual-motion machine when its brain is included in the accounting. As long as the demon never forgets its actions they can be undone; erasing its memory acts very much like the cycling of a piston in a steam engine. Rolf's colleague Charles Bennett was able to extend this intriguing connection between thought and energy to computing, showing that it is possible to build a computer that in theory uses as little energy as you wish, as long as you are willing to wait long enough to get a correct answer.

Rolf and Charles's work was seen as beautiful theory but remote from application until this decade, when power consumption suddenly became a very serious limit on computing. Supercomputers use so much power that it's hard to keep them from melting; so many PCs are going into office buildings that the electrical systems cannot handle the load; and portable computers need to be recharged much too often. As with steam engines and communication links, entropy provided a way to measure the energy efficiency of a computer and guide optimizations. Lower-power chips are now being designed based on these principles.

This history touches on fundamental theory (the microscopic explanation of macroscopic behavior), practical applications (efficient engines), far-reaching implications (digital communications), and even the nature of human experience (the physical limits on thinking). Yet which part is basic, and which is applied? Which led to which? The question can't be answered, and isn't particularly relevant. The evolution of a good idea rarely honors these distinctions.

The presumed pathway from basic research to applications operates at least as often in the opposite direction. I found out how to make molecular quantum computers only by trying to solve the short-term challenge of making cheap chips. If I had been protected to pursue my basic research agenda I never would have gotten there. And the converse of the common academic assumption that short-term applications should be kept at a distance from basic research is the industrial expectation that long-term research is remote from short-term applications.

A newly minted product manager once told me that he had no need for research because he was only making consumer electronics, even though his group wanted to build a Global Positioning

System (GPS) receiver into their product. GPS lets a portable device know exactly where it is in the world, so that for example a camera can keep track of where each picture was taken. It works by measuring the time it takes a signal to reach the receiver from a constellation of satellites. This requires synchronizing the clocks on all of the satellites to a billionth of a second. To measure time that precisely, each satellite uses an atomic clock that tells time by detecting the oscillation of an atom in a magnetic field. According to Einstein's theory of relativity, time slows down as things move faster, and as gravity strengthens. These effects are usually far too small to see in ordinary experience, but GPS measures time so finely that the relativistic corrections are essential to its operation. So what looks like a simple consumer appliance depends on our knowledge of the laws governing both the very smallest and largest parts of the universe.

Across this divide between the short and long term, a battle is now being fought over the desirability of basic versus applied research. The doers of research feel that they've honored their part of the bargain, producing a steady stream of solid results from transistor radios to nuclear bombs, and cannot understand why the continuing funding is not forthcoming. The users of research results find that the research community is pursuing increasingly obscure topics removed from their needs, and question the value of further support.

As a result both academic and industrial research are now struggling for funding. Many of the problems that inspired the creation of the postwar research establishment have gone away. The most fundamental physics experiments have been the studies at giant particle accelerators of the smallest structure of matter; these were funded by the Department of Energy rather than the National Science Foundation for an interesting historical reason.

After World War II the researchers who developed the nuclear bomb were clearly of great strategic value for the country, but few wanted to keep working for the military. So Oppenheimer made a deal for the government to fund particle physics research to keep the community together and the skills active, with the understanding that it might be tapped as needed for future military needs. The Department of Energy then grew out of the Atomic Energy Agency, the postwar entity set up to deal with all things nuclear. Now with the end of the cold war the demand for nuclear weapons is not what it used to be. At the same time, the particle experiments have reached a scale that new accelerators can no longer be afforded by a single country, if not a whole planet. Further scientific progress cannot come easily the way it used to, by moving to ever-higher energies to reach ever-finer scales. So there are now a lot of high-energy physicists looking for work.

Similarly, the original success of the transistor posed many research questions about how to make them smaller, faster, quieter. These have largely been solved; an inexpensive child's toy can now contain a chip with a million perfect transistors. There is still an army of physicists studying them, however, even though we understand just about everything there is to know about a transistor, and can make them do most everything we want them to.

What's happened is that disciplines have come to be defined by their domain of application, rather than their mode of inquiry. The same equations govern how an electron moves in a transistor and in a person. Studying them requires the same instruments. Yet when my lab started investigating the latter, to make furniture that can see and shoes that can communicate, along with all of the interest I was also asked whether it was real physics. This is a strange question. I can answer it, dissecting out the physics piece from the broader project. But I don't want to. I'd much rather be asked about what I learned, how it relates to what is known, and what its implications are.

Vannevar Bush's research model dates back to an era of unlimited faith in industrial automation, and it is just as dated. Factories were once designed around an assembly line, starting from raw materials and sequentially adding components until the product was complete. More recently it's been rediscovered that people, and machines, work better in flexible groups that can adapt to problems and opportunities without having to disrupt the work flow. Similarly, there is a sense in which a conveyor belt is thought to carry ideas from basic research to applied research to development to productization to manufacturing to marketing. The collision between bits and atoms is a disruption in this assembly line that cries out for a new way to organize inquiry.

Joe Paradiso turning an experiment into a magic trick, and Phil Rittmueller turning that into an auto safety product, provide a clue for where to look. A secret of how the Media Lab operates is traffic: there are three or four groups in the building every day. We're doing demos incessantly. This never happens in academia, because it is seen as an inappropriate intrusion into free inquiry. And it's not done in industry, where knowledge of internal projects is carefully limited to those with a need to know. Both sides divide a whole into pieces that add up to much less.

The companies can come in and tell us when we've done something useful, like the car seat. They can do that far better than we can. And, they can pose what they think are hard problems, which we may in fact know how to solve. Instead of writing grant proposals I see visitors, something that I would much rather do since I learn so much from them. After helping them with what I know now, I can use their support to work on problems that would be hard to justify to any kind of funding agency. Ted Adelson nicely captures this in a simple matrix:

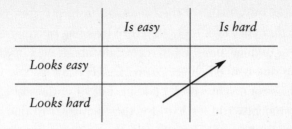

	Is easy	Is hard
Looks easy		
Looks hard		

Looks-easy-is-easy questions do not need research. *Looks-hard-is-hard* is the domain of grand challenges such as the space program that require enormous resources and very long-term commitments. *Looks-hard-is-easy* are the short-term problems that are easy to convey in a demo and to relate to an application. *Looks-easy-is-hard* are the elusive problems that really motivate us, like figuring out how machines can have common sense. It's difficult to explain why and how we do them, and hence is hard to show them in a demo. The lower-left corner pays the bills for the upper right, openly stealing from the rich projects to pay for the poor ones.

The third trick that makes this work is the treatment of intellectual property, the patents and copyrights that result from research. Academia and industry both usually seek to control them to wring out the maximum revenue. But instead of giving one sponsor sole rights to the result of one project, our sponsors trade exclusivity for royalty-free rights to everything we do. This lets us and them work together without worrying about who owns what. It's less of a sacrifice for us than it might sound, because very few inventions ever make much money from licensing intellectual property, but fights over intellectual property regularly make many people miserable. And it's less of a sacrifice for the sponsors than it might sound, because they leverage their investment in any one area with all of the other work going on. When they first arrive they're very concerned about protecting all of their secrets; once they notice that many of their secrets have preceded them and are already familiar to us, and that we can solve many of their internal problems, they relax and find that they get much more in return by being open.

The cost of their sponsorship is typically about the same for them as adding one new employee; there are very few people who can match the productivity of a building full of enthusiastic, unconstrained students, and it's hard for the companies to find and hire those people. Any one thing we do the sponsors can do better if they really want to, because they can throw much greater resources at a problem. The difficult thing is figuring out where to put those resources. We can happily learn from failing on nine projects, something that would bankrupt a company, and then hand off the tenth that succeeds.

This relationship in turn depends on another trick. We have a floor of people who wear suits and write memos and generally act corporate. This is essential to translate between the two cultures. Most every day I see industrial visitors asking to fund physics research, even though they don't realize that's what they are saying. They're posing problems without knowing where to look for solutions. It's unrealistic to expect them to come in understanding what disciplines are relevant, and how to work with them.

Sponsors as well as students need training. I realized this when a company visited, was excited by what they saw and became a sponsor, and just as quickly terminated their support, explaining that we were not delivering products in a timely fashion. The real fault lay not with them for so completely missing what a research lab is for, but with us for not recognizing and correcting their misunderstanding.

Over time, I've come to find that I like to work equally well with Nobel laureate scientists, virtuosic musicians, and good product managers. They share a remarkably similar sensibility. They have a great passion to create, bring enormous discipline to the task, and have a finely honed sense of what is good and what is bad.

While their approaches overlap, their training and professional experience do not. Mastery of one helps little with another. Yet this is exactly what companies presume when they look to their researchers to find applications for their work. I was visiting a large technology company once that was trying to direct its research division to be more useful. They did this by setting up a meeting that was more like a religious revival, where researchers were asked to come forward and testify with product plans. Not surprisingly, the researchers didn't have a clue. It takes an unusual combination of life experience and disposition to excel at perceiving market opportunities and relating them to research capabilities.

In between the successful scientist, artist, and manager is a broad middle ground where work does not answer to a research discipline, or a critical community, or the marketplace. Nicholas Negroponte and I spent a morning attending a research presentation at an industrial lab along with the company's chain of command. After a laborious discussion of their method, exactly how they went about designing and evaluating their project, they finally showed what they did. Nicholas was livid—an undergraduate could (and in fact did) accomplish the same thing in a few afternoons. He couldn't believe the waste of everyone's time. Amid all their process they had forgotten to actually *do* something. Because the same group of people were posing the question, doing the research, and evaluating the results, that message never reached them.

This is one more reason why demos are so important. They provide a grounding that is taken for granted in a mature field but that otherwise would be lacking in an emerging area. A steady stream of visitors offers a reality check that can help catch both bad and good ideas. This is a stretch for corporate cultures based around carefully controlling access, a protective reaction that can cause more harm than good in rapidly changing fields. I've seen corporate research labs that require management approval even for Internet access, limiting it to those employees who can show a job-related need to communicate. This distinction makes about as much sense as asking which employees have a job-related need to breathe; not surprisingly it keeps out much more information than it lets in.

Demos serve one more essential function. Our students spend so much time telling people what they're doing that they get good at communicating it to varied audiences, an essential skill lacking in most of science. And then eventually, after doing it long enough, they start to figure out for themselves what they're really doing. One of my former students told me that he found this to be the single most important part of his education, giving him an unfair advantage in his new lab because none of his colleagues had experience talking to nonscientists about their work.

Too much research organization is a matter of constraints, trying to decide who should do what. The sensible alternative that is regularly overlooked is contact, trying to bring together problems and solutions. Since the obvious connections have long since been found, this entails finding the nonobvious ones like the car seat that would not show up on a project list. The only successful way I've ever seen to do this is through regular visits that allow each side to meet the other.

It is a simple message with enormous implications. The result in the Media Lab is something that is not quite either traditional academia or industry, but that draws on the best of both. This model is not universally applicable; some problems simply take lots of time and money to solve. But it sure is fun, and it lets us pursue ideas without trying to distinguish between hardware and software, content and representation, research and application. Surprisingly often, working this way lets all sides get exactly what they want while doing just what they wish.

The inconvenient technology that we live with reflects the inconvenient institutional divisions that we live with. To get rid of the former, we need to eliminate the latter.

Information and Education

Nicholas Negroponte is the visible face of the Media Lab, an articulate advocate of being digital. Far fewer people know of Jerome Wiesner's role. Jerry was MIT's president, and Kennedy's science advisor. He's been described as the member of the administration during that exceptional era who was not only smart, but also wise.

As I started attending faculty meetings I soon discovered that Jerry was even more bored with them than I was. If I sat next to him he would tell me stories, like how the debate in the meeting we were ignoring mirrored a fight he had had in the Cabinet. Through these discussions I learned a secret: the Media Lab is really a front for an even more interesting project.

After a lifetime of shaping science, Jerry felt that there was a big hole in the middle of academia in general, and at MIT in particular. Disciplines were kept apart, basic and applied research happened in different places at different times, and industrial interaction was based on handing off results rather than intimate collaboration. Most serious of all, content of many kinds had no place on campus. As a physicist in a normal department I could look at transistors, but not toys. I might see industrialists, but not artists.

The Media Lab was his last grand project, a meta-experiment in organizing inquiry for a new era. He knew that this goal was so interesting and important that he could never really discuss it openly. There would be too many conflicting interests for it ever to be accomplished by a committee. Instead, Nicholas and his colleagues at the predecessor to the Media Lab, the Architecture Machine Group, provided the perfect research agenda and working style to create a laboratory laboratory.

It took me a long time to recognize this hidden project. When I started visiting, I didn't see the Media Lab as a place to do serious science; it was an entertaining diversion from the real work that I was doing elsewhere. I expected to eventually go to an industrial laboratory to set up my research group (I had found that I was too practical to be happy in a traditional academic setting). And Nicholas certainly didn't see the Media Lab as a place for something as dry and remote as a physics lab.

When we finally sat down to talk, he told me a story to explain his surprise at the prospect of my coming to the Media Lab. He said that if there was a monastery on top of a hill, and a brothel down in a valley, he wouldn't expect there to be too much traffic between the two institutions. I was struck by this image, but for the life of me I couldn't figure out which lab was which. I'm even less sure now.

The more we spoke, the more we realized that it did make sense for the Media Lab to have a physics group. For me, it would provide the support to work on physics in the new domains where it is needed. I knew how to do physics; the hard thing was getting access to the emerging context. For Nicholas, it would provide the tools and techniques to help open up computers and move information out into the world where it is needed.

In retrospect I'm embarrassed by how long it took me to decide that it was okay to have fun for a living. I had been trained to believe that the sorts of things I had been doing in the Media Lab, like working on Yo-Yo's cello, were done on the side by serious scientists, not as part of their day job. I thought that trying to do physics outside of a Physics department was going to be a radical and uncertain step.

I realized that it was going to work when students started showing up, explaining that they loved physics, but that they knew they did not want to study it in a traditional department. They were unwilling to disconnect their personal passions from their intellectual pursuits, and they did not want to follow a conventional academic career path with limited prospects for funding and employment. This was brought home one day by a headline in MIT's campus newspaper, "Cutbacks Announced in Dance and Physics."

I was also concerned about where my students would publish their research papers; I had expected that we would have to dissect out the pure academic nuggets from the impure context. Here again I realized that this was not going to be a problem when journal editors started showing up in my lab to solicit submissions. They recognize that the world is changing, and they don't want to keep publishing the same kinds of papers.

As my lab grew, my biggest surprise was watching how my students were reinventing the organization of their education. Scientific training has traditionally been based around extensive classwork, illustrated by occasional labs. My students were turning that inside out.

That they were in my lab at all was due to Edwin Land, the founder of Polaroid. He felt that MIT's undergrads should be in working laboratories participating in real research as it happened, rather than doing rote repetitions in lab classes long after the fact. He provided the seed funding for a very successful program to do just that.

The undergrads in my lab used it for far more than what Land originally envisioned. It became their home, the place where they learned to work with people and use disciplines to solve hard problems. Their classes took on a supporting role, providing the raw material that got shaped into an education in my lab. Over and over they told me that they had no idea why they were being taught something until they got to use it in one of our projects. For them, this experience was more like being in a studio than a classroom.

I found that one of the best predictors of a student's success working this way was their grades: I look to make sure they have a few F's. Students with perfect grades almost always don't work out, because it means they've spent their time trying to meticulously follow classroom instructions that are absent in the rest of the world. Students with A's and F's have a much better record, because they're able to do good work, and also set priorities for themselves. They're the ones most able to pose—and solve—problems that go far beyond anything I might assign to them.

The present system of classes does not serve the students, or their teachers, very well. One week an MIT undergrad, and independently an MIT professor, asked me the same question: how does the bandwidth of a telephone line relate to the bit rate of a modem? The bandwidth is the range of frequencies a phone line can pass, which is set by the phone companies and regulatory agencies. The bit rate is how fast data gets sent, all too noticeable as the speed with which a Web page loads. They were wondering how the bandwidth affects the bit rate that can be achieved, and what the prospects were for faster modems.

This is the problem that Shannon solved in the 1940s with his theory of information: the maximum possible bit rate is the bandwidth times the logarithm of one plus the ratio of the strength of the signal to the amount of noise. The bit rate can be improved by using more frequencies, by sending a stronger signal, or by decreasing the noise. Modems are now near the limit of a conventional phone channel, although the wires themselves can handle data much faster.

I was surprised to find that someone could be at MIT as long as both the student and professor had been, studying communications in many forms, and never have heard of a result as important as this. Not only that, both were unprepared to understand where it came from and what conclusions might be drawn from it.

Entropy shows up in two places in their question, in analyzing the noise that is intrinsic to the materials in a telephone, and in analyzing the information that can be carried by a message sent through the telephone. Although these two calculations are closely related, physicists learn how to do the former in one part of campus, and engineers the latter in another. Very few students manage to be sufficiently bilingual to be able to do both. Those who are either take so many classes beyond the usual load that they manage to squeeze in a few different simultaneous degrees, or take so few classes that they have enough time to put these pieces together on their own. Either path requires unusual initiative to answer such a reasonable question.

Faced with students who knew a lot about a little, I decided that I had to teach everything. This took the form of two semester-long courses, one covering the physical world outside of computers (The Physics of Information Technology), and the other the logical world inside computers (The Nature of Mathematical Modeling). My students loved them; some of my peers at MIT hated them. This was because each week I would cover material that usually takes a full semester.

I did this by teaching just those things that are remembered and used long after a class has ended, rather than everything that gets thrown in. This starts by introducing the language in an area so that the students, for example, know that bit rate and bandwidth are related by Shannon's channel capacity. Then I cover enough of each subject to enable students to be able to understand where the results come from and how they are used; at this point they would know how to calculate the channel capacity of a simple system.

Each week ends with pointers into the specialized literature so that, for example, students learn that Shannon's limit effectively can be exceeded by taking advantage of quirks of human perception. Although this brisk pace does a disservice to any one area, unless I teach this way most people won't see most things. It is only by violating the norms of what must be taught in each discipline that I can convey the value of the disciplines.

What connects the work in the Media Lab is a sensibility, a style of working, a set of shared questions and applications. It's not a discipline, a distinct body of knowledge that has stood the test of time and that brings order to a broad area of our experience. Progress on the former relies on the latter.

In many places computers are studied in separate departments, meaning that students who like computers get less exposure to mathematics and physics, and hence can have less insight into how computers work. This can result in their using the wrong tool for the wrong problem, and in their being unable to distinguish between hard things that look easy and easy things that look hard.

A number of computer scientists have told me that they want to mount display screens in glasses. They know that the eye cannot focus on something so close, so they intend to be clever and blur the image in just the right way to compensate for the optics of the eye. Unfortunately, this nice idea has no chance at all of working. The light passing through a lens must be specified by both its intensity and direction; the image on a display can control only the intensity. There is no pattern that can change the direction that light is emitted from a conventional display.

Or, some computer scientists were looking for a source of random numbers for cryptography, which depends for security on having a steady supply of random keys. They hit on the idea of aiming the camera connected to their computer at a 1960s lava lamp, and using the motion of the blobs as a source of randomness. In fact, a lava lamp is appealing to watch precisely because it is not completely random; there is a great deal of order in the motion. The electrical noise from a simple resistor is not only much easier to measure, it is one of the most random things that we know of. A far more convenient device can solve their problem far better.

One more computer scientist showed me the automated cart he developed for delivering mail. With great pride he turned it on and together we watched it crash into the wall. He was measuring the position of the cart by counting the revolutions of its wheels; if they slip at all it gets completely lost. I explained that there were techniques for measuring location inside a building, something like the GPS system used outside of a building. His instant response was, "Oh, that's hardware." It was the wrong level of description for him, so he was going to struggle on with software patches for a device that could never work reliably.

An education that forces people to specialize in hardware, or software, sends them out into the world with an erroneous impression that the two are easily separated. It is even embodied in our legal code, in the workings of the U.S. Patent Office. A patent must scrupulously distinguish between apparatus claims, on hardware, and method claims, on software. This means that "the medium is the message" is actually illegal: the message must be separated from the medium for patent protection.

The best patent examiners recognize that new technology is stretching the boundaries of old rules, and are flexible about interpreting them. The worst examiners refuse to accept that a single entity can simultaneously embody a physical apparatus and a logical method. I've spent years winding my way through the legal system with a particularly obtuse examiner who insists on trying to split an invention into its component hardware and software, even though the function of the device cannot be seen in either alone but arises only through their interaction.

Companies can't help but notice that these kinds of distinctions no longer make sense. I realized that we're living through a new industrial revolution when I saw how many senior executives visiting my lab said that they had no idea what business they were in. A newspaper company gathers and creates information, annotates and edits it, sells advertising, runs printing presses, and has a delivery fleet. As digital media make it possible to separate these functions, which of them are core competencies and which are legacy businesses? The answer is not clear, but the question certainly is.

One morning I met with a company that explained that they had reached the limit of complexity in what could be designed with a computer, and given the development cost of their product they needed better interfaces for designers to interact with large amounts of information. That afternoon a company explained that they had reached the limit of complexity in what could be designed with a computer, and given the development cost of their product they needed better interfaces for designers to interact with large amounts of information. The former made video games; the latter jet engines.

In the face of such rapid and uncertain change, a few lessons are emerging about how companies can respond. Instead of storing up large inventories of supplies and products, raw materials should arrive just in time to make products on demand. Management and computing should be done close to where the business happens, not in a central office. And production should be organized in flexible workgroups, not in regimented assembly lines or narrow departments.

Many of the same constraints apply to education, but few of these lessons have been learned. Universities go on filling students with an inventory of raw knowledge to be called on later; this is sensible if the world is changing slowly, but it is not. An alternative is just-in-time education, drawing on educational resources as needed in support of larger projects.

One of the best classes I ever taught at MIT was at one in the morning in Penn & Teller's warehouse in Las Vegas. We had gone out with a group of students to finish debugging the spirit chair prior to its first performance. In the middle of the night all of the hardware finally came together. Before the music could be tested, a computer had to combine the raw signals from the sensors to determine where the player's hands were. This is a classic data analysis problem. So I gave an impromptu lecture on function fitting to students who were eager (desperate, really) to learn it, and who then retained the ideas long after the performance was over.

The big new thing on campuses now is distance learning, using videoconferencing and videotapes to let people far from a place like MIT take classes there. This is an awful lot like mainframe computing, where there is a centralized learning processor to which remote learning peripherals get attached. Just like management or information processing, the most valuable learning is local. Far more interesting than letting people eavesdrop from a distance is providing them with tools to learn locally.

Some of this is happening naturally, as the Web lowers the threshold to make information widely available. More and more articles get shared through servers such as `http://xxx.lanl.gov` without needing to travel to research libraries to read expensive journals. Through the working notes my group puts on the Web, I discovered that we have virtual collaborators who use our results and provide thoughtful feedback long before the notes are formally published.

And some of this is happening through the tools. The same technology that lets us embed surprisingly sophisticated and flexible sensing and computing into children's toys or consumer appliances helps make the means for meaningful scientific experimentation much more widely accessible. A Lego set today can make measurements that were done only in specialized labs not too long ago.

Academics visiting the Media Lab usually start by asking what department we are in. They have a hard time understanding that the Media Lab is our academic home, not an appendage to another department. Uniquely at MIT, it both grants degrees and manages research. Learning and doing elsewhere are separated into departments and laboratories.

Military types visiting want to see an organizational chart, and are puzzled when we say that there isn't one. I spent a day with a Hollywood studio executive who kept asking who gets to greenlight projects. He was completely perplexed when I told him that more often than not projects get initiated by the undergrads. Most every day I come to work knowing exactly what we should do next, and most every day they show me I'm wrong.

The reality is that there are many different ways to view the organization of the Media Lab, all correct: by projects, by disciplines, by levels of description, by industrial consortia, by people. It functions as an intellectual work group that can easily be reconfigured to tackle new problems. Once visitors understand this (lack of) structure, the next question they ask is how the Media Lab can be copied, and why there aren't more already.

In part, the answer is that it just requires getting these few organizational lessons right, but few institutions are willing to make that leap. We've had an unblemished record of failure in helping start spin-offs from the Media Lab elsewhere. What always happens is that when it comes time to open the new lab, whoever is paying for it expects to run it, choosing the projects and controlling the intellectual property. They're unwilling to let go of either, letting the research agenda trickle up from the grassroots, and eliminating intellectual property as any kind of barrier to collaboration.

In part, the answer is that there are a few unique features of the environment of the Media Lab that don't show up on any formal documents. To start, running the Media Lab is a bit like driving a bumper car at an amusement park. While we're more or less in control, the rest of MIT provides a great deal of enabling physical and intellectual infrastructure. This leads to steady jostling, but it would be impossible for the Media Lab to function in isolation outside of that less-visible institutional support.

Even that has a hidden piece. It was only after spending time at both Harvard and MIT that I realized that Cambridge has one university with two campuses, one for technology and one for humanities. While of course there are plenty of exceptions, broadly MIT's strength as a technology school frees Harvard from even trying to lead the world in bridge building, and because of Harvard's strengths MIT can be less concerned about Classics. Since MIT does not have a school of education, or a film school, there is not a turf battle over the Media Lab working on those things.

Then there are the students. It's not that they're any smarter than students elsewhere; it's that they're so desperate to make things work. I've never seen any group of people anywhere so willing to go without eating and sleeping to reduce something to practice. It almost doesn't even matter what it is.

One day I came in and found that my lab had been co-opted to make a grass display, a lawn mounted on mechanical pixels that could move to make an ambient agricultural information channel. The students building it labored heroically to make the drive circuits that could handle that much power, and fabricate the structure to merge the drive solenoids with the sod. The only thing they couldn't do is tell me why they were doing it. Once they realized that it was possible, they could not conceive of not making one.

About a month later I came in and found that my students had been up many nights in a row, hacking an interface into a global satellite system to get sensor data off the summit of Mount Everest for an expedition that Mike Hawley was putting together. They developed, debugged, and deployed a satellite terminal in a few days, slept for a few days, then went back to whatever else they had been doing. No one had ever been able to collect weather data from on top of Everest, since most people there are just trying to stay alive. The data from their probe went via satellite to the Media Lab, and from there it went to planetary scientists as well as to base camp to be relayed up to climbers on the mountain.

When the designer of the satellite link, Matt Reynolds, showed up in my lab as a freshman he was already the best radiofrequency engineer I'd ever met, including during my time as a technician at Bell Labs. This is a technical specialty that few people really master, and usually then only after a lifetime of experience. It's hard not to believe that Matt's knowledge is genetic, or comes from a past life. He as much as confirmed that one evening after we attended a conference in San Francisco, when we were happily sitting on a deck in Big Sur watching a memorable sunset. I casually asked him how he chose to come to MIT. The conversation ground to a halt; it was almost as if I had said something inappropriate. He paused, struggled to answer, and then said that he had known he was going to go to MIT since he was a fetus. He didn't choose to go to MIT, he had a calling to serve.

Another freshman, Edward Boyden, appeared one day with a ten-page manifesto for the future of all research. It was the most interesting nonsense I've read; he didn't have any of the details right, but the spirit and sensibility were wise far beyond his limited scientific experience. He came in with almost no relevant skills. I set him to work helping a grad student; in January Edward learned computer programming, in February he learned how to do 3D graphics and digital video, in March how to numerically model the behavior of electric fields, so that in April he could put all the pieces together and render the physics underlying the sensors we were developing to see with fields. Having caught up to the grad student, Ed's only reaction was to be frustrated by his own slow pace.

It makes no sense to funnel that kind of raw problem-solving ability through a conventional curriculum. Edward learns too quickly, and is too intellectually omniverous, to be content following a prescribed path through knowledge. Matt spends a tiny fraction of each day taking care of the formal requirements of his classes, then gets back to learning things that go far beyond the classes. They arrived understanding a lesson that it usually takes students longer to learn: the goal of classes is to get you to not want to take classes. Once students know how to read the literature, and teach themselves, and find and learn from experts, then classes should take on a supporting role to help access unfamiliar areas.

One of the most meaningful moments in my life came in an airport lounge. I was waiting to fly back from my brother's wedding, and ended up sitting with our family rabbi. In this informal setting, with time before the plane and no distractions, I screwed up my courage to say something that I had wanted to tell him for a long time. I explained that I found the morality, and history, and teachings of Judaism to be deeply significant, but that I had a hard time reciting formal rituals I didn't believe in. He beamed, and said that the same was true for him. Letting me in on a secret, he told me that he saw many of the ceremonial aspects of Jewish observance as a formal scaffolding to engage people's attention while the real meaning of the religion got conveyed around the edges. The formal structure was valuable, but as a means rather than an end.

Rather than start with the presumption that all students need most of their time filled with ritual observance, the organization of the Media Lab starts by putting them in interesting environments that bring together challenging problems and relevant tools, and then draws on more traditional classes to support that enterprise. The faster the world changes, the more precious traditional disciplines become as reliable guides into unfamiliar terrain, but the less relevant they are as axes to organize inquiry.

Things That Think

In the beginning, our collective vision of computation was shaped by the reality: large machines, with lots of blinking lights, that were used by specialists doing rather ominous things for the military or industry. *Popular Mechanics* in 1949 made the bold guess that "Where a calculator on the ENIAC is equipped with 18,000 vacuum tubes and weighs 30 tons, computers in the future may have only 1,000 vacuum tubes and perhaps weigh 1½ tons." Later came the fictional images inspired by this reality, captured by the Jetsons' cartoon world that put the trappings of big computers everywhere, cheerfully filling their lives with the same kinds of buttons and blinking lights.

If we look around us now, the single most common reaction to computers was entirely missed by any of the historical visions: irritation. Computers taking over the world is not a pressing concern for most people. They're more worried about figuring out where the file they were editing has gone to, why their computer won't turn on, when a Web page will load, whether the battery will run out before they finish working, what number to call to find a live person to talk to for tech support.

The irritation can be more than petty. A 1997 wire story reported:

ISSAQUAH, Wash. (AP)—A 43-year-old man was coaxed out of his home by police after he pulled a gun on his personal computer and shot it several times, apparently in frustration.

Apparently? He shot it four times through the hard disk, once through the monitor. He was taken away for mental evaluation; they should have instead checked the computer for irrational and antisocial behavior.

There aren't many people left who want to live in *The Jetsons*'s world. I realized this during the early days of the World Wide Web, when I unsuccessfully searched the Net to find a picture of the Jetsons. People had made home pages for their cats, and dogs, and cars, but no one was moved to create one for this most technological of cartoons. Given all of the inconveniences of the Information Age, who would want to confront computers in still more places?

Over the last few years an alternative vision of the home or office of the future has been emerging, a rather retro one. In the research community, the Jetsons' use of information technology would be called "ubiquitous computing," making computing available anywhere and everywhere. I'm much more interested in unobtrusive computing, providing solutions to problems everywhere without having to attend to the computers. By bringing smarter technology closer to people it can finally disappear.

Instead of making room for mice, your furniture and floors can electromagnetically detect your gestures. Icons leave the screen and through embedded smart materials they merge with the tangible artifacts that we live with. A few bulky display screens get replaced with changeable electronic inks wherever we now paint or print. The local controllers for these things self-organize themselves into adaptive networks that don't break if any one element does. These data may go to and from an information furnace down in the basement, heating the bits for the whole house. This is a major appliance that handles the high-speed communication with the outside world, performs the computationally intensive tasks like rendering 3D graphics, and manages the enormous database of the accumulated experience of the household. Like any furnace it might need periodic maintenance, but when it's working properly it's not even noticed, delivering timely information through the information grates of the household.

A coffeemaker that has access to my bed, and my calendar, and my coffee cup, and my last few years of coffee consumption, can do a pretty good job of recognizing when I'm likely to come downstairs looking for a cup of coffee, without forcing me to program one more appliance. Although none of those steps represents a revolutionary insight into artifical intelligence, the result is the kind of sensible behavior that has been lacking in machines.

Marvin Minsky believes that the study of artificial intelligence failed to live up to its promise, not because of any lack of intelligence in the programs or the programmers, but because of the limited life experience of a computer that can't see, or hear, or move. A child has a wealth of knowledge about how the world works that provides the common sense so noticeably absent in computers. Similarly, Seymour Papert feels that the use of computers for education has gotten stuck. We learn by manipulating, not observing. It's only when the things around us can help teach us that learning can be woven into everyday experience. He's not looking to duplicate the mind of a good teacher; he just wants a tennis ball that knows how it has been hit so that it can give you feedback. Marvin and Seymour are looking for answers to some of the most challenging questions about improving technology, and the deepest questions about human experience, in the simplest of places. They believe that progress is going to come from creating large systems of interacting simple elements.

Just as people like Marvin and Seymour began to realize that, and people like Joe Jacobson and I began to discover that we could make furniture that could see or printing that could change, some unusual companies started showing up in the Media Lab. Steelcase was wondering whether your tabletop should be as smart as your laptop. Nike was thinking about the implications of the World Wide Web for footwear. This convergence of research interests, technological capabilities, and industrial applications led to the creation in 1995 of a new research consortium, called Things That Think (TTT).

The first decade of the Media Lab's life was devoted to the recognition that content transcends its physical representation. A story is much more than just ink on paper, or silver halide in celluloid film; once it is represented digitally then it's no longer necessary to create an artificial technical boundary between words and images, sights and sounds. The most important contribution from this era was iconoclasm. It was widely accepted then that it was the job of governments and industry alliances to fight over incompatible standards for new generations of television; now it's broadly accepted that the introduction of intelligence into the transmitter and receiver means that digital television can be scalable, so that the encoding can change if the goal is to send a little image to a portable screen or a giant image to a theater screen, and a program can bring a broader context with it, such as annotated commentary or connections to current information.

In its second decade, more and more of the work of the Media Lab revolves around the recognition that capable bits need capable atoms. The 150 or so industrial sponsors are loosely organized into three broad consortia. Walter Bender's News in the Future has content providers, asking how to search, filter, personalize, and distribute timely information. Andy Lippman's Digital Life is looking at what it means to live in a world of information, with questions of education, and identity, and entertainment. And TTT looks directly at how the physical world meets the logical world. These three groups can roughly be thought of as the bits, the people, and the atoms. They all need one another, but each provides a community to help focus on those domains. They overlap in focused groups at the center for areas like toys or cars.

TTT comprises forty companies broadly exploring intelligence everywhere but in traditional computers. The bottom layer that I direct is developing the materials and mechanisms to let objects sense and compute and communicate. The partners include technology companies like HP and Motorola considering new markets for their chips, and things makers like Nike and Steelcase. The middle level, run by Mike Hawley, looks at how to build systems out of these elements. If my shoe, and yours, is a computer, how do they find each other anywhere on the planet? Companies like Microsoft, AT&T, and Deutsche Telekom are interested in this kind of connectivity. At the top, guided by Tod Machover, are all of the capabilities enabled by a planet full of communicating footwear.

The application companies are asking mission-critical information technology questions that are wholly unmet right now, literally keeping the CEOs up at night worrying about how to address them. Moving computing from mainframes to the desktop was a trivial step; they need it out in the world where their business happens. Disney, for example, is interested in personalization. A theme park attraction can't respond appropriately if it knows only your average height and weight, not your language and gender. How can a one-cent ticket contain that information and be read from a distance? Federal Express has something of a mainframe model, sending all the packages to a central hub for processing. Just as the Internet introduced packet switching, routing chunks of data wherever they need to go, FedEx would like a package-switched network. How can a one-cent envelope route itself? Steelcase's customers want the file cabinets to find the file folders: how can a one-cent file folder communicate its contents? Becton Dickinson made a billion medical syringes last year that are sterile, sharp, and cost a penny. Diabetics are notoriously bad at monitoring their insulin intake; how can a smart syringe be made for a penny? Adding much more than that destroys the whole business model.

Health care is an interesting example. Right now it's really sick care, a system that gets invoked to fix problems rather than anticipate them. One of the biggest medical issues is compliance, getting people to do what's needed to keep them well. Billions of dollars are spent annually just taking care of people who didn't take their medicine, or took too much, or took the wrong kind. In a TTT world, the medicine cabinet could monitor the medicine consumption, the toilet could perform routine chemical analyses, both could be connected to the doctor to report aberrations, and to the pharmacy to order refills, delivered by FedEx (along with the milk ordered by the refrigerator and the washing machine's request for more soap). By making this kind of monitoring routine, better health care could be delivered as it is needed at a lower cost, and fewer people would need to be supervised in nursing homes, once again making headway on a hard problem by building interconnected systems of simple elements.

It's not possible to go to a technological consultancy and order up these kinds of solutions; they need new kinds of devices connected into new kinds of networks. The idea of things thinking is old. What's new is the necessity and possibility of solving these problems. That's the work of TTT.

TTT functions more as a "do tank" than a think tank. There are enormous social and industrial implications of the research, but these emerge more from understanding what's been accomplished than by trying to figure them out in advance. While the latter order is also valuable, we find that the former is better able to answer questions that we haven't even asked yet, and to fill needs that were not articulated because it was not possible to even conceive of a solution.

And the sponsors need partners to build a business in these emerging areas, but usually a great deal of negotiation is involved before they can work together. Just as we learn a lot from student projects that don't succeed, TTT provides a context for sponsor companies to try out new business models before the lawyers get involved.

The most immediate consequence of TTT for me has been how it has reshaped the Media Lab. When I came, most people typed at computers, and there were a few desultory machine tools sitting neglected in a corner. The Media Lab now has some of the best fabrication facilities of all kinds on the MIT campus. Even more striking, every group in the building now has an oscilloscope. This is the basic instrument used for developing and debugging electronics; its presence shows that people are designing and modifying their own hardware. Basic computer literacy is quickly coming to include machining, and circuit design, and microcontroller programming.

Very few people are left just sitting in front of a computer. Or are even left in the building; pursuing these ideas has entailed following them to other spaces. Not to home-of-the-future projects, which have generally been rather deadly affairs designed and inhabited by middle-aged white males in suburban settings, but to environments that reflect people's personal passions, whether a networked Volvo driving through Cambridge, or a smart space in the Smithsonian museum, or a probe on the summit of Mount Everest.

The biggest surprise is how quickly the work is progressing. When TTT started it was an elusive concept, a quirky domain out of most everyone's field of view. In just a few years it has zoomed into the mainstream, rapidly becoming almost as familiar as an earlier strange notion, multimedia. This pace in turn raises a few questions. The first is when TTT will "happen."

I don't expect there to be an epochal day when heavenly trumpets blare and TTT gets turned on. It's leaking out now, and like so many other technological revolutions, its real arrival will come when the question is no longer interesting. Many early products have been announced or are in the pipeline, and we've seen that people are very quick to incorporate these kinds of capabilities in how they live.

At one of the first TTT meetings, Mitch Resnick's group made smart name badges. These could be dipped into buckets containing answers to provocative questions, such as where you see society headed, or what you would read on a desert island. Then, when you met somebody, your badges would compare notes and light up a colored bar graph. A long green line meant that you agreed on everything, something that's nice to know. A short red or green line indicated a weak overlap. Most interesting was a long red line, showing that you disagreed on everything. These ended up being the most interesting ones to find, because you were guaranteed to have a lively discussion about anything. This kind of information was so useful that during the course of the meeting people incorporated it into the familiar gesture of shaking hands, squaring up and exposing their chests to make sure that the displays were easily seen. All of this happened without any discussion or explanation.

Another project, Tod Machover's Brain Opera, explored the implications of smart spaces for creating an opera that people enter into instead of watching from a distance. The audience shaped the composition through a room full of sensing, computing, communicating objects. At the premiere at Lincoln Center, Joe Paradiso, the technology director, came upon someone in great frustration pounding on a structural girder. Everything else in the environment had responded in interesting ways; this person had a hard time accepting that the girder just held up the building. As more and more things gain the ability to interact, a steel beam that is all body and no brain will indeed come to be seen as deficient.

The next question is what TTT will mean for the future of the Internet. Right now the Net is groaning under the explosion of new hosts and new users. Available bandwidth decreases as it gets shared among more and more people, and the routing of data becomes less and less reliable as it becomes harder to keep track of what is where. What will happen to these problems when the number of people using the Net is dwarfed by the number of things?

Not too much. A great deal of effort is going into upgrading the Internet to handle real-time audio and video, for the killer Internet application: telephony, or videoconferencing, or movies on demand (pick your favorite). But your toaster doesn't need to watch movies. Most of the devices being connected to the Net are bit dribblers, things that send a small amount of useful data rather than a continuous multimedia stream. These things easily fit into the margins around the heavy users of bandwidth.

If TTT can happen, then what about privacy? The most Orwellian fears for the future were not paranoid enough to worry about the very real possibility that even your eaves may be doing Big Brother's eavesdropping, or to wonder whose side your shoes will be on.

I believe that in the same way that bringing more technology closer to people is the way to make it disappear, it is also the path to protecting privacy. Cryptography is an arms race that the encoders can always win over the decoders. The effort to encrypt a message more securely grows far more slowly than the effort to crack the code. Because of this, the government has tried to mandate encryption schemes such as the Clipper chip that have official back doors that permit "authorized" agencies to decrypt them. These have failed, because software encryption is too easy to implement for it to be prevented. A favorite T-shirt at MIT contains a few lines of computer code implementing a widely known secure cryptosystem; the shirt is officially classified as a munition that cannot be exported from the country. Leaving the United States with this shirt, or even a laptop with encrypted passwords, violates the unenforceable law. The real limit to personal use of cryptography has not been regulatory, it's been inconvenience. Your personal information cannot be routinely protected unless the things around you know how to do that.

Even forgetting encryption, the official eavesdropping agencies are drowning in the torrent of data flowing through networks. A wiretap was originally just that, a wire attached on or near another one to pick up the signal. Setting one up merely required access to some part of the wire. That no longer works on a high-bandwidth digital network. It's hopeless to try to pick one phone call out of a billion bits per second passing through an optical fiber. This means that legal eavesdroppers are reduced to pleading for convenient access to the data. Their demand took the form of the Communications Assistance for Law Enforcement Act (CALEA), passed in 1994, mandating manufacturers of telephone switching equipment to install ports for legal government surveillance of calls passing through the switches.

Like cryptography, this is a battle that the eavesdroppers can't win. It's far easier to generate data than it is to analyze it. As the network bandwidth increases, as the number of nodes connected to the network increases, and as efficient use of the network replaces easily intercepted messages with signals that appear to be random, the technological challenges presented to the eavesdroppers are going to surpass even what can be solved by legislation. The answer to Big Brother may well be little brother, a world full of communicating things providing a Lilliputian counter to the central control of information.

In the end I think that the strength of these technical arguments is going to turn privacy from a regulatory issue to a matter of personal choice. Right now the price of your insurance is based on your crude demographics, rather than your personal details. If you drive safely, and you let your car insurance company have access to data from your car that confirmed that, then you could pay less than the bozo who cuts you off. If you eat well, and you're willing to let your life insurance company talk to your kitchen, then you could be rewarded for having a salad instead of a cigarette. The insurance company would not be in the business of enforcing any morality; they would be pricing the expected real cost of behavior. These kinds of communication can be encrypted without revealing irrelevant but sensitive identifying information, such as where you're driving or when you're home. And the ability to turn insurance from a static document into an on-line tool does not mean that you have to participate; complete privacy will remain available, but it will cost more because the insurance company has less to go on to price your policy.

It's as irritating to lug along at 55 mph in the middle of the Arizona desert as it is to be passed at 155 mph on a winding rain-slicked German autobahn. Setting speed limits is a political process that necessarily results in an unhappy compromise between competing interests and varying needs. Driving faster carries a cost in pollution, road wear and tear, and safety. All of these things have qualifiers, such as the straightness of the road, the condition of your car, and your driving experience. If you're willing to share this information with the road then speed limits can be dynamically chosen and contain useful information, rather than the current practice of setting universally ignored rates based on the lowest common denominator. Privacy alone is not an absolute good; what matters is making a sensible trade-off between private and shared interests. Connecting things provides a way for these trade-offs to become matters of personal rather than national policy.

A final question about TTT is what it implies for the stratification of society, locally and globally. If the haves are already diverging from the have-nots, what's going to happen when there are all these new things to have?

In thinking about this essential question, it's important to recognize that technology rarely cause or fixes social problems; those are a consequence of much larger social forces. In 1747 James Lind, a British ship surgeon, figured out that British sailors were dying at sea from scurvy because they didn't have fresh fruit (containing vitamin C). Packing limes let the ships stay at sea much longer, projecting the might of the British Navy much farther. Was that advance in nutrition good or bad? Sailors' lives were saved, other lives were lost from the wars they then fought, lives were saved from the order imposed by the ensuing empire, more lives were lost getting rid of the empire. It's too simple to credit or blame that history on vitamin C.

Granting that, there are modest grounds to think that TTT can help rather than exacerbate the existing divisions. There's a new foundation growing out of the Media Lab, called 2B1, to develop and deploy information technology for children in developing countries. It's almost, but not quite, a Media Lab Peace Corps. Instead of drilling wells for water, 2B1 will provide bit wells for information.

I was struck by many things when a large group of representatives of developing countries came to visit the Media Lab for the first time. I expected there to be a great deal of sensitivity about cultural imperialism, or cultural pollution, in presuming to connect African villages to the Internet. Instead, in many different ways, people expressed the belief that the world is changing quickly, and it is far more elitist to insist that developing countries progress through all of the stages of the Industrial Revolution before they're allowed to browse the Web. They want to participate on an equal footing with the rest of the world. The loss of local culture is a valid concern, but then the Web has also served to foster the creation of community among nearby people who had not had a means to communicate among themselves before its arrival.

The technological challenges they've presented line up almost perfectly with the work of TTT. A computer in the middle of the veldt must need no spare parts, has to run without an electrical outlet or telephone line, can't require calls to technical support, must be usable without a manual, and be insanely cheap and indestructible. Such a thing is not in the direct lineage of a desktop PC; it requires a range of new technologies.

Life in the veldt already provides a model for what this might eventually look like; biology is the master designer of adaptive systems that fix themselves. Extrapolating the history of the Media Lab from bits to atoms leads rather directly to biology.

For the foreseeable future we've got a lot of work to do to create around people inanimate technologies that they can trust. Beyond that, it's impossible not to speculate about implanting them in people. Your body is already the ultimate wearable computer; why not move the heads-up display to your retina? Striking work is being done in augmenting people's sensory and motor disabilities in this manner, but I don't trust anyone yet to be able to do these things on a purely discretionary basis. It's not until the technologies outside of people become so useful and reliable that I couldn't live without them that I would consider trying to live with them. I'm not ready to see a dialog box reading, "You just crashed, OK?" and then have to take a nap to reboot.

If the present work of TTT succeeds, implants are the natural next step, equally intriguing and frightening. Even more so is what could come after that, editing the genome so that you grow the right parts. The genome already guides the fabrication of such exquisite machines as the eye and the hand (as well as more problematical hacks like the knee). What's so privileged about our current eye design? We now do know a lot about optics, and chemistry, and could design eyes that have a broader spectral response, say, or that could look backward as well as forward. There are no longer serious ontological debates about the design of the eye as proof of the existence of a God, but we haven't taken seriously the converse that if the design of the eye does not represent divine intervention, and we don't intend to replace one deity with another by deifying evolution, then the eye is open to mortal improvement. And who says we just have to upgrade our existing senses. Growing up I was disappointed when I realized that it appeared that I didn't have ESP, but I do know how to use a cell phone to talk around the world. How about adding radios to brains?

The attendant ethical, social, and scientific challenges are staggering, but the point stands that our current construction represents one evolved and evolving solution and that it is open to improvement. Evolution is a consequence of interaction, and information technology is profoundly changing how we interact; therefore, it's not crazy to think about an impact on evolution. If I'm far from being ready to let someone implant a chip, I'm certainly nowhere near being willing even to entertain seriously a discussion of aftermarket additions to the genome, but I have to admit that that is the logical destination of current trends. I fear, and hope, that we eventually reach that point. You'll be able to tell we're getting close when the Media Lab starts hiring molecular biologists.

In between the dimly glimpsed, terrifying, and thrilling possibility of an evolution in evolution, and the present reality of an over-hyped digital revolution, lies a foreseeable future that is increasingly clear and appealing. In retrospect it looks like the rapid growth of the World Wide Web may have been just the trigger charge that is now setting off the real explosion, as things start to use the Net so that people don't need to. As information technology grows out of its awkward adolescence, bringing more capabilities closer to people is proving to be the path to make it less obtrusive and more useful. The implications of this change are on display in the laboratory now, and given the combination of industrial push and consumer pull it's hard to believe that they will not soon be equally familiar elsewhere.

Now that media has become multi, and reality has become virtual, and space is cyber, perhaps we can return to appreciating the rest of reality. As a technologist so frequently annoyed by technology, as an academic developing products, as a member of the species wondering about the evolution of the species, I can't imagine a more exciting mission than merging the best of the world that we are born into with that of the worlds we're creating.

Afterword

Reading about science is like reading about food or exercise. The description can be very interesting, but it's no substitute for the real thing. That applies to this book. In explaining what is happening I've made no attempt to convey or document the details. The best way to follow up is to learn something of those details.

A good starting point is the Web site for my research group, http://www.media.mit.edu/physics, and for the Things That Think consortium, http://www.media.mit.edu/ttt. These have papers on many of the projects I've mentioned, as well as video clips showing them in action and articles about them.

Then come my two texts, *The Nature of Mathematical Modeling* and *The Physics of Information Technology,* both published by Cambridge University Press. These are self-contained treatments on the worlds inside and outside of computers, and include all of the technical material and supporting references behind this book. Although the texts reach a graduate level, my hope and experience has been that they're useful for the motivated but otherwise inexperienced reader in learning to speak the technical language and join the discussion.

Index